WASTEWATER TREATMENT FUNDAMENTALS III

ADVANCED TREATMENT

OPERATOR CERTIFICATION STUDY QUESTIONS

2022

Water Environment Federation
601 Wythe Street
Alexandria, VA 22314–1994 USA
http://www.wef.org

Water Professionals International
9400 Plum Drive, Suite 160
Urbandale, IA 50022
http://www.gowpi.org

Copyright © 2022 by the Water Environment Federation. All Rights Reserved. Permission to copy must be obtained from WEF.

ISBN: 978-1-57278-445-1

Water Environment Research, *WEF*, and *WEFTEC* are registered trademarks of the Water Environment Federation.

IMPORTANT NOTICE

This peer-reviewed publication is intended to provide information through a review of technical practices and detailed procedures that research and experience have shown to be functional and practical.

The material presented in this publication has been prepared in accordance with generally recognized engineering principles and practices and is for general information only. This information should not be used without first securing competent advice from an industry professional with respect to its suitability for any general or specific application.

No reference made in this publication to any specific method, product, process, or service constitutes or implies an endorsement, recommendation, or warranty thereof by WEF.

WEF makes no representation or warranty of any kind, whether expressed or implied, concerning the accuracy, product, or process discussed in this publication and assumes no liability.

Anyone using this information assumes all liability arising from such use, including but not limited to infringement of any patent or patents.

WEF works hard to ensure that the information in this publication is accurate and complete. However, it is possible that the information may change after publication, and/or that errors or omissions may occur. We welcome your suggestions on how to improve this publication and correct errors.WEF disclaims all liability for, damages of any kind arising out of use, reference to, or reliance on information included in this publication to the full extent provided by state and Federal law.

About WEF

The Water Environment Federation (WEF) is a not-for-profit technical and educational organization of 30,000 individual members and 75 affiliated Member Associations representing water quality professionals around the world. Since 1928, WEF and its members have protected public health and the environment. As a global water sector leader, our mission is to connect water professionals; enrich the expertise of water professionals; increase the awareness of the impact and value of water; and provide a platform for water sector innovation. To learn more, visit www.wef.org.

About WPI

Water and wastewater treatment professionals are on the front lines of protecting public health and the environment. Since 1972, Water Professionals International (formerly known as the Association of Boards of Certification) has been the central water industry authority that ensures these women and men are prepared to meet the standards that their communities can trust in through our testing and certification services.

WPI includes 100+ certifying authorities representing more than 40 states, 10 Canadian provinces and territories, and several international and tribal programs. WPI has assisted in the certification of more than 500,000 water and wastewater treatment professionals, including those who have earned a WPI Certification or Professional Operator (PO) Certification.

Prepared by

Chapter 1 Industrial Wastewater and Pretreatment

Alison Ling

Sara Arabi, PhD, PE, BCEE

Carlos D. Claros

Tyson Schlect, PE

Tat Ebihara

Hemanth Haft

Chapter 2 Advanced Safety Considerations

Nicole Zerbel Ivers

Tim Page-Bottorff

Megan Yoo Schnieder, PE

Chapter 3 Physical and Chemical Treatment

Raj Chavan

Kamrun Ahmadi

Kristin O'Neill, Brown and Caldwell

Chapter 4 Advanced Activated Sludge

John C. Kabouris, PhD, PE, ENV SP, Stantec

Thomas Nogaj, PhD, PE, Stantec

Parsa Pezeshk

Chapter 5 Nontraditional Disinfection

Gary Hunter, PE, BCEE, ENVSP

Leonard W. Casson, PhD, PE, BCEE, ENVSP

Bernadette Drouhard

Isaiah Spencer-Williams

Chapter 6 Membranes

Kerry J. Howe, PhD, PE, BCEE, Howe Water Science LLC

Elise Chen, PE, Trussell Technologies, Inc.

Rodrigo Tackaert, PE, Trussell Technologies, Inc.

Aleksey N. Pisarenko, PhD, Trussell Technologies, Inc.

Chapter 7 Water Reuse

Anya Kaufmann, PE, Trussell Technologies, Inc.

Rodrigo Tackaert, PE, Trussell Technologies, Inc.

Eileen Idica, PhD, PE, Trussell Technologies, Inc.

Aleksey N. Pisarenko, PhD, Trussell Technologies, Inc.

Chapter 8 Characterization and Sampling of Sludge

Mahmudul Hasan

Adrian Romero

Arifur Rahman, PhD, PE

Chapter 9 Management of Solids

Kari Brisolara

Rachel Watson

Keerthisaranya Palanisamy

Chapter 10 Additional Stabilization Methods

Anthony Tartaglione

James Tiberius Schiera, PE

Jay S. Kemp

Jorj Long

Chapter 11 Odor Control

Chriso Petropoulou

Mike Hickey

Jesse Varsho

Chapter 12 Instrumentation

Sidney Innerebner, PhD, PE, CWP, PO, WEF Fellow, Indigo Water Group

Stacy J. Passaro, PE, WEF Fellow, Passaro Engineering, LLC

EURING Oliver Grievson, CEng, CEnv, CSci, CWEM, FCIWEM, FIEnvSc, FIET, FInstMC, Z-Tech Control Systems

Melody White

Chapter 13 Supervisory Control and Data Acquisition (SCADA) Systems

Bob Daly

Gerald Di Noia

Chapter 14 Leadership and Management

Tricia H. Kilgore, PE, Beaufort-Jasper Water & Sewer Authority

Melissa Darr, PE

Dan O'Sullivan

Ryan Lacharity, MASc, PEng, Region of Waterloo

Christel Dyer

Aaron J. Rivard, Genesee County Drain Commissioner: Water & Waste Services

With assistance and review provided by

Chandra Baker

Vanessa Borkowski, PE

Chein-Chi Chang, PhD, PE, Chang Tech
International

James P. Cooper, PE

Frank DeOrio, PO

Houssam B. Eljerdi, Chemical Engineer, PE

Richard Finger

Val S. Frenkel, PhD, PE, DWRE, GREELEY and
HANSEN

Ronald Gehr

Georgine Grissop, PE, BCEE

Vaughan Harshman, PE

Marialena Hatzigeorgiou, PE

Sidney Innerebner, PhD, PE, CWP, PO, WEF
Fellow, Indigo Water Group

Matthew Jalbert

Jeannette Klamm

Dale Kocarek

Anna Lacourt, MEng, PEng

(Emy) Wenxin Liu, PhD, PE

Yanjin Liu, PhD, PE

Jennifer Loudon

Indra Maharjan, Ontario Clean Water Agency

Karthik Manchala

Andy O'Neil, PO

Stacy J. Passaro, PE, WEF Fellow, Passaro
Engineering, LLC

Natalia Perez, PE, New York City Department
of Environmental Protection (NYCDEP)

Lance Salerno

Frederick Tack, PE, D.WRE

Steve Walker, CWP

Under the direction of the **Technical Practice Committee**

Andrew R. Shaw, PhD, PE, *Chair*

Jon Davis, *Vice-Chair*

Dan Medina, PhD, PE, *Past Chair*

H. Azam, PhD, PE

G. Baldwin, PE, BCEE

S. Basu, PhD, PE, BCEE, MBA

P. Block, PhD

C.-C. Chang, PhD, PE

R. Chavan, PhD, PE, PMP

A. Deines

M. DeVuono, PE, CPESC, LEED AP BD+C

N. Dons, PE

T. Dupuis, PE

T. Gellner, PE

C. Gish

V. Harshman, PE

G. Heath, PE

M. Hines

M. Johnson

N.J.R. Kraakman, Ir., CEng.

J. Loudon

C. Maher

M. Mulcare

C. Muller, PhD, PE

T. Page-Bottorff

A. Rahman, PhD, PE

J. Reina

V. Sundaram, PhD, PE

M. Tam, PE

E. Toot-Levy

Contents

Water Professionals International Formulas

Wastewater Treatment, Collection, Industrial Waste, & Wastewater Laboratory Exams

*Pie wheel format for this equation is shown at the end of the formulas.

Alkalinity, mg/L as $CaCO_3 = \dfrac{\text{(Titrant Volume, mL)(Acid Normality)(50 000)}}{\text{Sample Volume, mL}}$

$Amps = \dfrac{\text{Volts}}{\text{Ohms}}$

Area of Circle* = (0.785)(Diameter^2)

Area of Circle = (3.14)(Radius^2)

Area of Cone (lateral area) = (3.14)(Radius)$\sqrt{\text{Radius}^2 + \text{Height}^2}$

Area of Cone (total surface area) = (3.14)(Radius)(Radius + $\sqrt{\text{Radius}^2 + \text{Height}^2}$)

Area of Cylinder (total exterior surface area) = [End #1 SA] + [End #2 SA] + [(3.14)(Diameter)(Height or Depth)]
    Where SA = surface area

Area of Rectangle* = (Length)(Width)

Area of Right Triangle* = $\dfrac{\text{(Base)(Height)}}{2}$

Average (arithmetic mean) = $\dfrac{\text{Sum of All Terms}}{\text{Number of Terms}}$

Average (geometric mean) = $[(X_1)(X_2)(X_3)(X_4)(X_n)]^{1/n}$ The *nth* root of the product of *n* numbers

Biochemical Oxygen Demand (seeded), mg/L = $\dfrac{[\text{(Initial DO, mg/L)} - \text{(Final DO, mg/L)} - \text{(Seed Correction, mg/L)}][300 \text{ mL}]}{\text{Sample Volume, mL}}$

Biochemical Oxygen Demand (unseeded), mg/L = $\dfrac{[\text{(Initial DO, mg/L)} - \text{(Final DO, mg/L)}][300 \text{ mL}]}{\text{Sample Volume, mL}}$

Blending or Three Normal Equation = $(C_1 \times V_1) + (C_2 \times V_2) = (C_3 \times V_3)$ *Where $V_1 + V_2 = V_3$; C = concentration, V = volume or flow; Concentration units must match; Volume units must match*

CFU/100 mL = $\dfrac{[(\text{# of Colonies on Plate})(100)]}{\text{Sample Volume, mL}}$

Chemical Feed Pump Setting, % Stroke = $\dfrac{\text{Desired Flow}}{\text{Maximum Flow}} \times 100\%$

Chemical Feed Pump Setting, mL/min = $\dfrac{\text{(Flow, mgd)(Dose, mg/L)(3.785 L/gal)(1 000 000 gal/mil. gal)}}{\text{(Feed Chemical Density, mg/mL)(Active Chemical, \% express as a decimal)(1440 min/d)}}$

Chemical Feed Pump Setting, mL/min = $\dfrac{\text{(Flow, m}^3\text{/d)(Dose, mg/L)}}{\text{(Feed Chemical Density, g/cm}^3\text{)(Active Chemical, \% express as a decimal)(1440 min/d)}}$

Circumference of Circle = (3.14)(Diameter)

$$\text{Composite Sample Single Portion} = \frac{(\text{Instantaneous Flow})(\text{Total Sample Volume})}{(\text{Number of Portions})(\text{Average Flow})}$$

$$\text{Cycle Time, min} = \frac{\text{Storage Volume, gal}}{(\text{Pump Capacity, gpm}) - (\text{Wet Well Inflow, gpm})}$$

$$\text{Cycle Time, min} = \frac{\text{Storage Volume, m}^3}{(\text{Pump Capacity, m}^3/\text{min}) - (\text{Wet Well Inflow, m}^3/\text{min})}$$

$$\text{Degrees Celsius} = \frac{(^\circ\text{F} - 32)}{1.8}$$

Degrees Fahrenheit = (°C)(1.8) + 32

$$\text{Detention Time} = \frac{\text{Volume}}{\text{Flow}} \qquad \textit{Units must be compatible}$$

Dilution or Two Normal Equation = $(C_1 \times V_1) = (C_2 \times V_2)$ *Where C = Concentration, V = volume or flow; Concentration units must match; Volume units must match*

Electromotive Force, V* = (Current, A)(Resistance, ohm - Ω)

$$\text{Feed Rate, lb/d*} = \frac{(\text{Dosage, mg/L})(\text{Flow, mgd})(8.34 \text{ lb/gal})}{\text{Purity, \% expressed as a decimal}}$$

$$\text{Feed Rate, kg/d*} = \frac{(\text{Dosage, mg/L})(\text{Flowrate, m}^3/\text{d})}{(\text{Purity, \% expressed as a decimal})(1000)}$$

$$\text{Filter Backwash Rate, gpm/sq ft} = \frac{\text{Flow, gpm}}{\text{Filter Area, sq ft}}$$

$$\text{Filter Backwash Rate, L/(m}^2{\cdot}\text{s)} = \frac{\text{Flow, L/s}}{\text{Filter Area, m}^2}$$

$$\text{Filter Backwash Rise Rate, in./min} = \frac{(\text{Backwash Rate, gpm/sq ft})(12 \text{ in./ft})}{7.48 \text{ gal/cu ft}}$$

$$\text{Filter Backwash Rise Rate, cm/min} = \frac{\text{Water Rise, cm}}{\text{Time, min}}$$

$$\text{Filter Yield, lb/sq ft/hr} = \frac{(\text{Solids Loading, lb/d})(\text{Recovery, \% expressed as a decimal})}{(\text{Filter Operation, hr/d})(\text{Area, sq ft})}$$

$$\text{Filter Yield, kg/m}^2{\cdot}\text{h} = \frac{(\text{Solids Concentration, \% expressed as a decimal})(\text{Sludge Feed Rate, L/h})(10)}{(\text{Surface Area of Filter, m}^2)}$$

Flowrate, cu ft/sec* = (Area, sq ft)(Velocity, ft/sec)

Flowrate, m³/sec* = (Area, m²)(Velocity, m/s)

$$\text{Food-to-Microorganism Ratio} = \frac{\text{BOD}_5, \text{lb/d}}{\text{MLVSS, lb}}$$

$$\text{Food-to-Microorganism Ratio} = \frac{\text{BOD}_5, \text{kg/d}}{\text{MLVSS, kg}}$$

Force, lb* = (Pressure, psi)(Area, sq in.)

Force, newtons* = (Pressure, Pa)(Area, m²)

$$\text{Hardness, as mg CaCO}_3/\text{L} = \frac{(\text{Titrant Volume, mL})(1000)}{\text{Sample Volume, mL}}$$

Only when the titration factor is 1.00 of ethylenediaminetetraacetic acid (EDTA)

$$\text{Horsepower, Brake, hp} = \frac{(\text{Flow, gpm})(\text{Head, ft})}{(3960)(\text{Pump Efficiency, \% expressed as a decimal})}$$

$$\text{Horsepower, Brake, kW} = \frac{(9.8)(\text{Flow, m}^3/\text{s})(\text{Head, m})}{(\text{Pump Efficiency, \% expressed as a decimal})}$$

$$\text{Horsepower, Motor, hp} = \frac{(\text{Flow, gpm})(\text{Head, ft})}{(3960)(\text{Pump Efficiency, \% expressed as a decimal})(\text{Motor Efficiency, \% expressed as a decimal})}$$

$$\text{Horsepower, Motor, kW} = \frac{(9.8)(\text{Flow, m}^3/\text{s})(\text{Head, m})}{(\text{Pump Efficiency, \% expressed as a decimal})(\text{Motor Efficiency, \% expressed as a decimal})}$$

$$\text{Horsepower, Water, hp} = \frac{(\text{Flow, gpm})(\text{Head, ft})}{3960}$$

Horsepower, Water, kW = (9.8)(Flow, m³/s)(Head, m)

$$\text{Hydraulic Loading Rate, gpd/sq ft} = \frac{\text{Total Flow Applied, gpd}}{\text{Area, sq ft}}$$

$$\text{Hydraulic Loading Rate, m}^3/(\text{m}^2\cdot\text{d}) = \frac{\text{Total Flow Applied, m}^3/\text{d}}{\text{Area, m}^2}$$

Loading Rate, lb/d* = (Flow, mgd)(Concentration, mg/L)(8.34 lb/gal)

$$\text{Loading Rate, kg/d*} = \frac{(\text{Flow, m}^3/\text{d})(\text{Concentration, mg/L})}{1000}$$

Mass, lb* = (Volume, mil. gal)(Concentration, mg/L)(8.34 lb/gal)

$$\text{Mass, kg*} = \frac{(\text{Volume, m}^3)(\text{Concentration, mg/L})}{1000}$$

$$\text{Mean Cell Residence Time or Solids Retention Time, days} = \frac{(\text{Aeration Tank TSS, lb}) + (\text{Clarifier TSS, lb})}{(\text{TSS Wasted, lb/d}) + (\text{Effluent TSS, lb/d})}$$

Milliequivalent = (mL)(Normality)

$$\text{Molarity} = \frac{\text{Moles of Solute}}{\text{Liters of Solution}}$$

$$\text{Motor Efficiency, \%} = \frac{\text{Brake hp}}{\text{Motor hp}} \times 100\%$$

$$\text{Normality} = \frac{\text{Number of Equivalent Weights of Solute}}{\text{Liters of Solution}}$$

$$\text{Number of Equivalent Weights} = \frac{\text{Total Weight}}{\text{Equivalent Weight}}$$

$$\text{Number of Moles} = \frac{\text{Total Weight}}{\text{Molecular Weight}}$$

$$\text{Organic Loading Rate-RBC, lb SBOD}_5/1000 \text{ sq ft/d} = \frac{\text{Organic Load, lb SBOD}_5/\text{d}}{\text{Surface Area of Media, 1000 sq ft}}$$

$$\text{Organic Loading Rate-RBC, kg SBOD}_5/\text{m}^2\cdot\text{d} = \frac{\text{Organic Load, kg SBOD}_5/\text{d}}{\text{Surface Area of Media, m}^2}$$

$$\text{Organic Loading Rate-Trickling Filter, lb BOD}_5/1000 \text{ cu ft/d} = \frac{\text{Organic Load, lb BOD}_5/\text{d}}{\text{Volume, 1000 cu ft}}$$

$$\text{Organic Loading Rate-Trickling Filter, kg/m}^3\cdot\text{d} = \frac{\text{Organic Load, kg BOD}_5/\text{d}}{\text{Volume, m}^3}$$

$$\text{Oxygen Uptake Rate or Oxygen Consumption Rate, mg/L}\cdot\text{min} = \frac{\text{Oxygen Usage, mg/L}}{\text{Time, min}}$$

$$\text{Population Equivalent, Organic} = \frac{(\text{Flow, mgd})(\text{BOD, mg/L})(8.34 \text{ lb/gal})}{0.17 \text{ lb BOD/d/person}}$$

$$\text{Population Equivalent, Organic} = \frac{(\text{Flow, m}^3/\text{d})(\text{BOD, mg/L})}{(1000)(0.077 \text{ kg BOD/d}\cdot\text{person})}$$

$$\text{Power, kW} = \frac{(\text{Flow, L/s})(\text{Head, m})(9.8)}{1000}$$

$$\text{Recirculation Ratio-Trickling Filter} = \frac{\text{Recirculated Flow}}{\text{Primary Effluent Flow}}$$

$$\text{Reduction of Volatile Solids, \%} = \left(\frac{\text{VS in} - \text{VS out}}{\text{VS in} - (\text{VS in} \times \text{VS out})}\right) \times 100\% \qquad \textit{All information (In and Out) must be in decimal form}$$

$$\text{Removal, \%} = \left(\frac{\text{In} - \text{Out}}{\text{In}}\right) \times 100\%$$

$$\text{Return Rate, \%} = \frac{\text{Return Flowrate}}{\text{Influent Flowrate}} \times 100\%$$

$$\text{Return Sludge Rate-Solids Balance, mgd} = \frac{(\text{MLSS, mg/L})(\text{Flowrate, mgd})}{(\text{RAS Suspended Solids, mg/L}) - (\text{MLSS, mg/L})}$$

$$\text{Slope, \%} = \frac{\text{Drop or Rise}}{\text{Distance}} \times 100\%$$

$$\text{Sludge Density Index} = \frac{100}{\text{SVI}}$$

$$\text{Sludge Volume Index, mL/g} = \frac{(\text{SSV}_{30}, \text{mL/L})(1000 \text{ mg/g})}{\text{MLSS, mg/L}}$$

$$\text{Solids, mg/L} = \frac{(\text{Dry Solids, g})(1\,000\,000)}{\text{Sample Volume, mL}}$$

$$\text{Solids Capture, \% (Centrifuges)} = \left[\frac{\text{Cake TS, \%}}{\text{Feed Sludge TS, \%}}\right] \times \left[\frac{(\text{Feed Sludge TS, \%}) - (\text{Centrate TSS, \%})}{(\text{Cake TS, \%}) - (\text{Centrate TSS, \%})}\right] \times 100\%$$

Solids Concentration, mg/L $= \dfrac{\text{Weight, mg}}{\text{Volume, L}}$

Solids Loading Rate, lb/sq ft/d $= \dfrac{\text{Solids Applied, lb/d}}{\text{Surface Area, sq ft}}$

Solids Loading Rate, kg/m²·d $= \dfrac{\text{Solids Applied, kg/d}}{\text{Surface Area, m}^2}$

Solids Retention Time: *see Mean Cell Residence Time*

Specific Gravity $= \dfrac{\text{Specific Weight of Substance, lb/gal}}{8.34 \text{ lb/gal}}$

Specific Gravity $= \dfrac{\text{Specific Weight of Substance, kg/L}}{1.0 \text{ kg/L}}$

Specific Oxygen Uptake Rate or Respiration Rate, (mg/g)/h $= \dfrac{(\text{OUR, mg/L} \cdot \text{min})(60 \text{ min})}{(\text{MLVSS, g/L})(1 \text{ h})}$

Surface Loading Rate or Surface Overflow Rate, gpd/sq ft $= \dfrac{\text{Flow, gpd}}{\text{Area, sq ft}}$

Surface Loading Rate or Surface Overflow Rate, L/m²·d $= \dfrac{\text{Flow, L/d}}{\text{Area, m}^2}$

Total Solids, % $= \dfrac{(\text{Dried Weight, g}) - (\text{Tare Weight, g})}{(\text{Wet Weight, g}) - (\text{Tare Weight, g})} \times 100\%$

Velocity, ft/sec $= \dfrac{\text{Flowrate, cu ft/s}}{\text{Area, sq ft}}$

Velocity, ft/sec $= \dfrac{\text{Distance, ft}}{\text{Time, sec}}$

Velocity, m/s $= \dfrac{\text{Flowrate, m}^3\text{/s}}{\text{Area, m}^2}$

Velocity, m/s $= \dfrac{\text{Distance, m}}{\text{Time, s}}$

Volatile Solids, % $= \left[\dfrac{(\text{Dry Solids, g}) - (\text{Fixed Solids, g})}{(\text{Dry Solids, g})}\right] \times 100\%$

Volume of Cone* $= (1/3)(0.785)(\text{Diameter}^2)(\text{Height})$

Volume of Cylinder* $= (0.785)(\text{Diameter}^2)(\text{Height})$

Volume of Rectangular Tank* $= (\text{Length})(\text{Width})(\text{Height})$

Water Use, gpcd $= \dfrac{\text{Volume of Water Produced, gpd}}{\text{Population}}$

Water Use, L/cap·d $= \dfrac{\text{Volume of Water Produced, L/d}}{\text{Population}}$

Watts (AC circuit) = (Volts)(Amps)(Power Factor)

Watts (DC circuit) = (Volts)(Amps)

$$\text{Weir Overflow Rate, gpd/ft} = \frac{\text{Flow, gpd}}{\text{Weir Length, ft}}$$

$$\text{Weir Overflow Rate, L/m·d} = \frac{\text{Flow, L/d}}{\text{Weir Length, m}}$$

$$\text{Wire-to-Water Efficiency, \%} = \frac{\text{Water hp}}{\text{Motor hp}} \times 100\%$$

$$\text{Wire-to-Water Efficiency, \%} = \frac{(\text{Flow, gpm})(\text{Total Dynamic Head, ft})(0.746 \text{ kW/hp})(100\%)}{(3960)(\text{Electrical Demand, kW})}$$

Pie Wheels

- To find the quantity above the horizontal line: multiply the pie wedges below the line together.
- To solve for one of the pie wedges below the horizontal line: cover that pie wedge, then divide the remaining pie wedge(s) into the quantity above the horizontal line.
- Given units must match the units shown in the pie wheel.
- When US and metric units or values differ, the metric is shown in parentheses, e.g. (m^2).

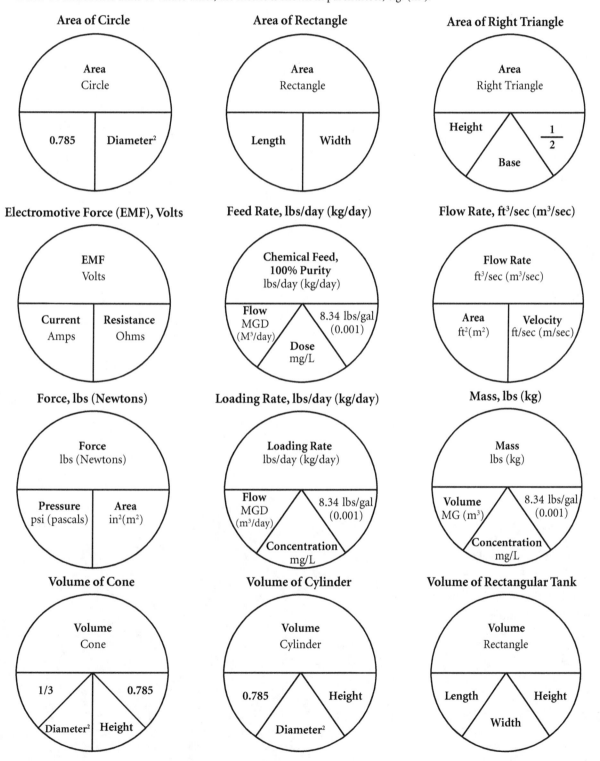

Conversion Factors

1 ac = 4046.9 m² or 43 560 sq ft

1 ac ft of water = 326 000 gal

1 atm = 33.9 ft of water

   = 10.3 m of water

   = 14.7 psi

   = 101.3 kPa

1 cfs = 0.646 mgd

   = 448.8 gpm

1 cu ft of water = 7.48 gal

     = 62.4 lb

1 ft = 0.305 m

1 ft H_2O = 0.433 psi

1 gal (US) = 3.79 L

    = 8.34 lb of water

1 gr/gal (US) = 17.1 mg/L

1 ha = 10 000 m²

1 hp = 0.746 kW

   = 746 W

   = 33 000 ft lb/min

1 in. = 25.4 mm or 2.54 cm

1 L/s = 0.0864 ML/d

1 lb = 0.454 kg

1 m of water = 9.8 kPa

1 m² = 1.19 sq yd

1 m³ = 1000 kg

   = 1000 L

   = 264 gal

1 metric ton = 2205 lb

1 mile = 5280 ft

1 mgd = 694 gpm

   = 1.55 cfs

   = 3.785 ML/d

Population equivalent (PE), hydraulic = 378.5 L/cap·d

            = 100 gpd/cap

PE, organic = 0.077 kg BOD/cap·d

    = 0.17 lb BOD/cap/d

1 psi = 2.31 ft of water

   = 6.89 kPa

1 ton = 2000 lb

1% = 10 000 mg/L

π or pi = 3.14

CHAPTER 1

Industrial Wastewater and Pretreatment

1. The Federal Pretreatment Act defines industrial and commercial users differently from one another.
   - ☐ True
   - ☐ False

2. What two parameters are most often considered when calculating pretreatment surcharges for industrial users?
   - a. COD and BOD
   - b. BOD and TSS
   - c. TSS and TDS
   - d. TOC and BOD

3. A pass-through pollutant reporting to a WRRF from an industrial source can affect its ability to meet permit limits for that pollutant.
   - ☐ True
   - ☐ False

4. Pretreatment of industrial wastes is regulated under the NPDES program.
   - ☐ True
   - ☐ False

5. Pretreatment permitting can only be done by the local WRRF receiving industrial wastes.
   - ☐ True
   - ☐ False

6. Which of the following is not included in pretreatment standards and requirements?
   - a. General and specific prohibitions
   - b. Source control measures
   - c. Categorical pretreatment standards
   - d. Local limits

7. All WRRFs are required to establish local pretreatment programs.
   - ☐ True
   - ☐ False

8. Generally, a WRRF acts as the pretreatment "control authority" and the state or U.S. EPA acts as the pretreatment "approval authority."
   - ☐ True
   - ☐ False

9. Pretreatment control authorities must establish an Enforcement Response Plan (ERP), which includes a framework for which of the following:
   - a. Procedures to investigate and respond to industrial user noncompliance
   - b. Procedures to develop local limits
   - c. Procedures to report industrial user noncompliance to the U.S. EPA

10. Which of the following criteria is NOT used to define a significant industrial user (SIU) under National Pretreatment Regulations?
    a. An industrial user that discharges an average of 95 m³/d (25 000 gpd) or more of process wastewater to the WRRF
    b. An industrial user that contributes process wastewater making up 5% or more of the average dry weather hydraulic or organic capacity of the WRRF
    c. An industrial user subject to federal categorical pretreatment standards
    d. An industrial user that produces more than 10 000 tonnes (11 000 tons) of hazardous waste per month

11. A local limit supersedes the categorical pretreatment standard for a given pollutant if it is more stringent.
    ☐ True
    ☐ False

12. Maximum Allowable Headworks Loading (MAHL) is a method used to establish local limits based on how much of a specific pollutant the WRRF can accept from all permitted industrial users and still ensure protection against pass-through and interference.
    ☐ True
    ☐ False

13. The most common parameters measured in industrial discharges are pH, BOD, and TSS.
    ☐ True
    ☐ False

14. Time proportional sampling is recommended for industrial discharges where possible.
    ☐ True
    ☐ False

15. 40 CFR 403 specifies different reporting requirements for categorical industrial users (CIUs) versus SIUs.
    ☐ True
    ☐ False

16. The difference between the local limit and categorical pretreatment standard is where the criteria are applicable (point of discharge or after treatment).
    ☐ True
    ☐ False

17. Water resource recovery facilities are required to inspect all industrial dischargers at least once a year.
    ☐ True
    ☐ False

18. Per- and polyfluoroalkyl substances (PFAS) have established regulatory limits at both the state and federal level.
    ☐ True
    ☐ False

19. Industrial facilities cannot directly discharge wastewater to a surface water via an NPDES permit; they must discharge to a WRRF.
    ☐ True
    ☐ False

20. Contaminants of emerging concern (CECs) are called that instead of "emerging contaminants" because many have been present for decades but are only now being subjected to increased regulatory and public health scrutiny as measurement methods improve and we understand more about their toxicity.
    ☐ True
    ☐ False

21. Treating organic CECs is similar, regardless of their specific chemistry.
    ☐ True
    ☐ False

22. The effects of polymers used in industrial wastewater processes are neutralized by the activated sludge at the WRRF and do not affect settleability.
    ☐ True
    ☐ False

23. All food processing wastewater is high in BOD and TSS but low in nutrients.
   ☐ True
   ☐ False

24. A high-volume cooling water discharge to a WRRF may be a concern if
   a. Thermal load could affect WRRF biological treatment performance.
   b. Water volume could increase WRRF receiving water temperature.
   c. Either a or b

25. Quaternary ammonium compounds (QACs) discharged by industrial users typically affect downstream WRRF processes by inhibiting bacteria in the activated sludge process.
   ☐ True
   ☐ False

26. Polymer flocculants typically affect downstream WRRF processes by inhibiting bacteria in the activated sludge process.
   ☐ True
   ☐ False

27. Landfill leachate discharged to WRRFs often results in nutrient overloading, which can cause the WRRF to exceed nutrient permit limits.
   ☐ True
   ☐ False

28. Landfill leachate discharges can have a negative effect on WRRF processes, even if it makes up less than 5% of the total WRRF influent flow.
   ☐ True
   ☐ False

29. Equalization at the industrial site refers to upstream industrial processes meant to control the amount of wastewater produced.
   ☐ True
   ☐ False

30. Clean-in-place (CIP) processes at industrial sites typically use acids, bases, or organic chemicals to clean equipment.
   ☐ True
   ☐ False

31. Wastewater discharges from the food and beverage sector do not exhibit this characteristic:
   a. High salt concentrations
   b. High temperatures
   c. High concentrations of organic compounds
   d. High heavy metals concentrations

32. Bench testing can be a good option for WRRFs to evaluate alternative chemical options and doses to counter interferences from industrial discharges.
   ☐ True
   ☐ False

33. Flotation treatment systems such as dissolved air flotation (DAF) work by adding fine gas bubbles to float solids and FOG to the surface.
   ☐ True
   ☐ False

34. High-strength wastewater is wastewater that contains high amounts of dissolved salts or TDS.
   ☐ True
   ☐ False

35. Pretreatment of wastewater with heavy metals is effective because the heavy metal concentration is diluted before treatment.
   ☐ True
   ☐ False

36. Heavy metals can accumulate in WRRF solid residuals like waste activated sludge and aerobic and anaerobic digesters.
    - ☐ True
    - ☐ False

37. Why might a dairy plant wastewater discharger to a WRRF pretreat high-strength BOD wastewater?
    a. Reduce toxicity of BOD to WRRF activated sludge treatment
    b. Reduce BOD surcharge costs
    c. Reduce TSS surcharge costs
    d. Both b and c

38. Why would heavy metal-laden wastewaters be pretreated before discharge to a WRRF?
    a. Reduce toxicity
    b. Minimize effect on biosolids quality
    c. Both a and b

39. Ion exchange is typically the initial pretreatment process used to remove metals.
    - ☐ True
    - ☐ False

CHAPTER 2

Advanced Safety Considerations

Safety Programs/Incident Prevention

1. The main goal of a safety and health program is to avoid citations from OSHA.
   - ☐ True
   - ☐ False

2. A facility health and safety committee can be a way to actively participate in a safety program, provide feedback, and discuss opportunities for improvement.
   - ☐ True
   - ☐ False

3. The following is true about reward and recognition programs:
   a. They encourage worker participation and positive safety behaviors.
   b. They are important to achieving low injury rates.
   c. They should always be in the form of money or large prizes.
   d. They are the same as an "incentive."

4. Hazard identification is _____.
   a. A permit to perform work where the permit is signed by the most senior operator
   b. Not important in a health and safety program because risks are commonly known and expected
   c. A process used to evaluate if any particular situation, item, thing, and so on may have the potential to cause harm
   d. The safety professional's responsibility to identify and then notify the other workers in the facility

Hierarchy of Controls

1. The last line of defense is PPE. Barriers and procedures should come first.
   - ☐ True
   - ☐ False

2. An example of substitution is
   a. Installing guardrails around a tank
   b. Using a water-based paint instead of solvent-based
   c. Wearing a hard hat and safety goggles
   d. Putting a guard on a pump

3. There are five levels in the hierarchy of controls.
   - ☐ True
   - ☐ False

Personal Protective Equipment

1. The basic forms of PPE recommended for most WRRF operators may include the following:
   a. Long pants, long sleeves, gloves, and respirator
   b. Goggles, gloves, and protective clothing
   c. Hard hat, protective clothing, and eye protection
   d. Hard hat, protective clothing, and eye and hearing protection

2. Choose the best glove for handling diesel fuel:
   a. Neoprene
   b. Latex
   c. Butyl
   d. Nitrile

Fall Protection and Prevention

1. OSHA requires training for fall protection hazards.
   ☐ True
   ☐ False

2. Fall protection is required where a walking or working surface has an unprotected side or edge that is how many meters (feet) or more above a lower level?
   a. 1.8 m (6 ft)
   b. 3 m (10 ft)
   c. 7.6 m (25 ft)
   d. 1.2 m (4 ft)

3. Falls are one of the leading causes of injuries and deaths on the job.
   ☐ True
   ☐ False

4. A passive fall protection system includes which of the following?
   a. Fall harness
   b. Lanyard
   c. Guardrail
   d. Anchor point

5. Fall-restraint equipment consists of a harness and a lanyard that is tethered to a substantial anchor point.
   ☐ True
   ☐ False

6. Fixed ladders must be equipped with a personal fall arrest system, ladder safety system, cage, or well when it exceeds how many meters (feet)?
   a. 1.2 m (4 ft)
   b. 1.8 m (6 ft)
   c. 7.3 m (24 ft)
   d. 9.1 m (30 ft)

Lockout/Tagout (Control of Hazardous Energy)

1. Lockout/tagout devices can be removed by another employee for the following reasons:
   a. If they have permission from the employee who applied the device.
   b. If they are authorized to do so by the person in charge of the LOTO program.
   c. If the employee who put on the lock is on lunch.
   d. If the work is completed.

2. Which of the following is true about energy control procedures?
   a. They cover maintenance schedules.
   b. They include periodic inspections.
   c. They do not need to include OSHA requirements.
   d. Energy control procedures are optional.

3. Lockout/tagout guidance is only a general recommendation and is not legally required by OSHA.
   ☐ True
   ☐ False

Confined Space Entry

1. The U.S. Occupational Safety and Health Administration has defined a *confined space* using three conditional criteria: (1) large enough and employee can enter, (2) limited means of access or egress, and (3) not designed for continuous human occupancy.
   ☐ True
   ☐ False

2. Permit required confined spaces can be declassified as non-permitted confined spaces.
   ☐ True
   ☐ False

3. An entrant needs to know the hazards before entering a confined space.
   ☐ True
   ☐ False

4. Which of the following is a responsibility of the attendant?
   a. Communicate with the human resources department.
   b. Know the weather forecast.
   c. Know the signs or symptoms of potential exposure by the entrant.
   d. Terminate and cancel entry permit.

5. Hazardous atmospheres include
   a. Nitrogen deficiency
   b. Combustible atmospheres
   c. Nontoxic chemicals or gases
   d. Dirt and dust

Hazard Communication

1. Hazard communication is also known as the
   a. The employer's right to refuse
   b. The employer's right to know
   c. The employee's right to know standard
   d. None of the above

2. What are the three broad elements to a written HAZCOM program?
   a. Labeling, SDSs, and employee information and training
   b. Science, math, and technology
   c. U.S. Occupational Safety and Health Administration needs, employer needs, and chemical manufacturer needs
   d. There are no requirements for a written program

3. SDS is an acronym for
   a. Safety down the street
   b. Safely describe sewers
   c. Safety data sheet
   d. Sewers destroy safety

Chemical Handling, Storage, and Response

1. Which chemical spill can be handled by a single individual?
    a. Incidental
    b. Operational
    c. Emergency response
    d. None of the above

2. How many hours of training are needed at the operational level?
    a. 2 hours
    b. 4 hours
    c. 8 hours
    d. 24 hours

3. Who should pay for required equipment, PPE, and training?
    a. The employee
    b. The employer
    c. The community
    d. The federal government

4. What does the acronym HAZWOPER mean?
    a. Hazardous Wastes Operations Team
    b. Hazardous Waste Operations and Emergency Response
    c. Hazardous Worker Personnel
    d. Hazardous Work Operations

Heavy Equipment and Machinery

1. Guards come standard on all machines and equipment.
    ☐ True
    ☐ False

2. What type of crane is shown in this photograph?

    a. Tower crane
    b. Mobile crane
    c. Bridge crane
    d. Gantry crane

3. A MEWP or lift that exceeds the load rating and is operated on a slope, grade, or ramp could tip over.
    ☐ True
    ☐ False

4. Who requires guards on equipment?
    a. ANSI
    b. Individual WRRFs
    c. OSHA
    d. The state

5. Working in close proximity to energized power lines or other electrically energized conductors can increase risk of electrocution.
    ☐ True
    ☐ False

6. The rated load of the crane should be plainly marked on each side of each crane.
    ☐ True
    ☐ False

7. It is only required to inspect cranes and hoists one time per year.
    ☐ True
    ☐ False

8. Label the parts of the forklift with the following terms:

    Fork   Mast   Overhead Guard   Counterweight   Tilt Cylinder   Load Backrest

9. Cranes can be modified and load capacity rerated by someone who knows how to operate the crane.
    ☐ True
    ☐ False

10. Forklift training only includes classroom training; it's not required to conduct a practical test and be evaluated on the ability to operate.
    ☐ True
    ☐ False

11. It is acceptable for operators/occupants of MEWPs to achieve additional height in the platform by climbing on the guardrail or using ladders.
    ☐ True
    ☐ False

CHAPTER 3
Physical and Chemical Treatment

Chemical Dosing

1. The adhesion of atoms, ions, or molecules to a surface is called
   a. Adsorption
   b. Absorption
   c. Molecule merging
   d. Fusion

2. What should you never mix with chlorine?
   a. Ammonia
   b. Ferric chloride
   c. Sulfuric acid
   d. Activated carbon

3. What is a typical dose range for aluminum sulfate in a jar test?
   a. 20 to 40 mg/L
   b. 0 to 5 mg/L
   c. 5 to 20 mg/L
   d. 40 to 50 mg/L

4. What is an optimal pH range for aluminum sulfate coagulation?
   a. 0–1
   b. 1–3
   c. 5–6
   d. 6–7

5. What is compatible with sulfuric acid?
   a. Steel
   b. Water
   c. Sulfur
   d. PVC

6. Which of the following *are* compatible?
   a. Alum, copper
   b. Ammonia, chlorine
   c. Sulfuric acid, NaOH
   d. Alum, Type 316 SS

7. High turbidity is observed post-settling or post-filtration. What is the most likely cause?
   a. High alkalinity
   b. Proper chemical dose
   c. Clog or tear in chemical feed tubing
   d. A pH of 6.5

8. What would cause a chemical feed tubing line to rupture at a location away from the peristaltic pumps?

    a. Pump operating at speeds that are too high

    b. Coagulant or acid has dissolved the tubing

    c. Pump has been started against an open valve

    d. Pump has been started against a closed valve

9. What is a limitation of the jar test?

    a. Lack of mixing power

    b. Insufficient coagulant dose

    c. Lack of ability to mimic full-scale chemical dosing lag

    d. Jars are too small

Coagulation and Flocculation

1. What is the primary goal of coagulation and flocculation?

    a. To adjust the alkalinity of the wastewater

    b. To create biosolids

    c. To stabilize the colloidal particles

    d. To remove turbidity, bacteria, and nutrients

2. Colloidal particles and natural organic matter are typically positively charged.

    ☐ True

    ☐ False

3. Which is a type of flocculator?

    a. A high-speed vertical turbine pump

    b. A rapid mixer

    c. A baffled concrete basin

    d. Peristaltic pump

4. Hydraulic basins can create

    a. Large flocs

    b. Turbidity

    c. Whirlpools

    d. Solids

5. Coagulation (rapid mix) requires a high

    a. Velocity gradient (G)

    b. Detention time

    c. Turbidity

    d. Chlorine demand

6. Flocculation requires a

    a. High-velocity gradient

    b. Low detention time (t)

    c. High detention time (t)

    d. High power input

7. Flocculation requires a

    a. Low coagulant dose

    b. High velocity gradient (G)

    c. Low velocity gradient (G)

    d. Low detention time (t)

8. During eutrophication, the water becomes
    a. Clear
    b. Oxygen sufficient
    c. Oxygen deficient
    d. Muddy

9. Floc breakage occurs at
    a. Low chemical dose
    b. High chemical dose
    c. High velocity gradient
    d. Low velocity gradient

10. Paddle wheels have a large
    a. Rotational speed range
    b. Mixing area
    c. Power consumption
    d. Footprint

Clarification

1. Which of the following is a type of clarifier basin?
    a. Alabama flocculator
    b. Serpentine flow
    c. Upflow
    d. Anoxic basin

2. In plug flow, the velocity of the fluid is assumed to be _____ across any cross-section of pipe perpendicular to the axis of the pipe.
    a. Constant
    b. Varied
    c. Supercritical
    d. Subcritical

3. When used as part of an advanced water treatment train (coagulation–flocculation-clarification) what is a primary difference from clarifiers used for primary or secondary treatment in conventional wastewater treatment?
    a. Overflow rate
    b. Detention time
    c. Sludge return and recycle rate and removal

4. The following is a major design parameter for settling tanks:
    a. Overflow rate
    b. Alkalinity
    c. Sludge wasting rate
    d. Coagulant concentration

5. Short-circuiting is an issue with clarification tanks where wind blows across the water surface, reducing net volume and detention time in the tank.
    ☐ True
    ☐ False

6. When a constant water temperature is not ensured in clarification tanks, thermal stratification may occur with cold "dead" zones on the bottom of the tank.
    ☐ True
    ☐ False

7. When poor floc formation is observed, what should be done to determine if it is due to inadequate chemical dosing?
    a. Run a jar test
    b. A timed cone settling test
    c. Alkalinity and pH measurement
    d. Sludge density test

8. Calculate the overflow rate of a rectangular clarifier with a side water depth of 4.572 m (15 ft), length of 18.3 m (60 ft), width of 3.66 m (12 ft), and flow of 7570.8 m³/d (2 mgd):
    a. 73.15 m/d (240 ft³/sq ft/day)
    b. 7.56 m/d (24.8 ft³/sq ft/day)
    c. 372 m³/m²·d (1220 ft/day)
    d. 34.56 m³/m²·d (113.39 ft/day)

9. Which is a common problem in a settling tank?
    a. Poor floc formation
    b. Floc breakup and excessive turbulence
    c. Low sludge formation
    d. Inadequate mixing

10. In an advanced wastewater clarifier as part of a coagulation–flocculation treatment train, why would you not want to waste all of the sludge?
    a. You do want to turn up the pumps and waste all of the sludge.
    b. Biomass is required in the treatment train to seed the basins in secondary treatment.
    c. Previously precipitated particles in the return sludge may form a nucleus for more rapid flocculation.
    d. It costs too much power to waste all the sludge.

Filtration

1. A mechanism of removal during filtration is
    a. Straining
    b. Settling
    c. Fusion
    d. Destabilization

2. Direct filtration is best used when turbidity is less than
    a. 50
    b. 100
    c. 10
    d. 15

3. Conventional filtration (with coagulation, flocculation, clarification) is best used in water with _____ turbidity levels.
    a. Low
    b. High

4. What would not cause the filter headloss to be low?
    a. Inadequate dosage of polymers as filter aid
    b. Coagulant feed system malfunction
    c. Change in coagulant demand
    d. A sudden increase in alkalinity

5. If the percentage of backwash water recycled is consistently high (over 5%) for a filter system, the operator should aim to
    a. Improve solids settling
    b. Increase filter aid dosage
    c. Add more aluminum sulfate
    d. Adjust the pH

6. Label the three zones (A, B, and C) on the following chart graph of a filter run:

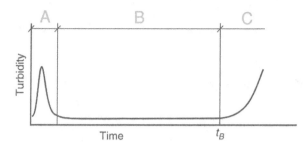

7. An operator observes high headloss through the filter bed, they should
    a. Adjust the coagulant dose
    b. Increase flocculation speed
    c. Decrease the sludge wasting rate
    d. Remove the filter from service and backwash

8. Air bubble accumulation within the filter (and subsequent increasing headloss) is most likely caused by
    a. Malfunctioning diffusers
    b. Excessive coagulant dose
    c. Not enough coagulant dose
    d. Dissolved oxygen saturated filter influent

9. At your facility you have noticed that your filter becomes more difficult to clean during hot summer months and requires more backwash water. This is most likely due to
    a. Excessive algae growth during warm weather
    b. Excessive nutrient discharge from organic runoff during warm weather
    c. A decrease in flowrates due to higher water consumption during summer months
    d. Decreased water viscosity due to higher temperatures

Granular Activated Carbon

1. Granular activated carbon is regenerated through what mechanism?
    a. Heat
    b. Lime slurry
    c. Digestion
    d. Replacement

2. The pores in GAC particles _____ the surface area of the particle.
    a. Decrease
    b. Agitate
    c. Increase
    d. Destroy

3. Granular activated carbon columns in series _____ total column height.
    a. Increase
    b. Decrease
    c. Double
    d. Halve

4. Granular activated carbon adsorption is typically used as a _____ treatment process.
    a. Primary
    b. Secondary
    c. Tertiary
    d. Preliminary

5. What is the most common GAC contactor configuration?
   a. Uplow
   b. Downflow
   c. Vortex
   d. Braided

6. When particles do not adsorb and make their way through the GAC system, this is called _____.
   a. Pushing
   b. Breakthrough
   c. Non-adsorption
   d. Bleeding

7. The expected performance of a GAC system is completely dependent on what?
   a. The upstream treatment processes
   b. The downstream treatment processes
   c. The disinfection treatment process
   d. The screening process

8. Upflow contactor may also be called what?
   a. Expanded bed
   b. Lifted bed
   c. Downward bed
   d. Raised bed

9. Kilns and multiple hearth furnace equipment are associated with what process in a GAC system?
   a. Regeneration
   b. Backwashing
   c. Testing
   d. Liberation

10. Granular activated carbon systems are effective at removing _____ organics.
    a. Floating
    b. Dissolved
    c. Wet
    d. Particulate

Ion Exchange

1. What type of resin is considered naturally occurring?
   a. Polymorphic
   b. Zeolite
   c. Cationic
   d. Anionic

2. What is the most common material used for ionic exchange regeneration?
   a. Lime slurry
   b. Hydroxide
   c. Polymer
   d. Carbon slurry

3. An anionic exchanger transfers what type of ion?
   a. Positive
   b. Negative
   c. Neutral
   d. Double

4. A strong acid reactor operates best at what pH?
    a. 7
    b. 9
    c. 2
    d. 10

5. The operation capacity is always _____ the exchange capacity.
    a. More than
    b. Less than
    c. Equal to
    d. Subtracted from

6. What step of operations follows regeneration?
    a. Backwashing
    b. Rinsing
    c. Service
    d. Cleaning

7. What is the indication that a backwash needs to occur?
    a. Breakthrough
    b. Temperature change
    c. pH change
    d. Media color change

8. Which of the following applications is ion exchange commonly used for to reduce hardness?
    a. Softening
    b. Demineralization
    c. Decarbonization
    d. Dissociation

9. Empty bed contact time is typically measured in what units?
    a. Minutes
    b. Hours
    c. Million gallons per day
    d. Gallons per minute

10. A resin's selectivity is typically indicated by what symbol?
    a. J
    b. K
    c. Q
    d. M

CHAPTER 4
Advanced Activated Sludge

Microbiology

1. Where should a microscopic exam be performed?
   a. Aeration basin
   b. Clarifiers
   c. Return activated sludge (RAS)
   d. Waste activated sludge (WAS)

2. Focusing a microscope is accomplished by changing the position of the body tube with respect to the position of the stage.
   ☐ True
   ☐ False

3. The resolving power, or resolution of an optical system, is a measure of how many detectable colors can be seen with the microscope.
   ☐ True
   ☐ False

4. Phase contrast microscopy is of value in the study of opaque microorganisms and is routinely used for examining organisms like rotifers and nematodes that cause bulking.
   ☐ True
   ☐ False

5. The objective and ocular lenses of a microscope are situated side-by-side, and the objective lens is always located under the specimen.
   ☐ True
   ☐ False

6. What are the four focus points in performing a microscopic exam?

7. One micron meter (μm) equals 1/1000 of a meter or 1/10 of a millimeter.
   ☐ True
   ☐ False

8. One of the primary differences between the low-magnification stereoscopic microscope and the high-magnification compound microscope is that the stereoscopic microscope typically uses UV light and the compound microscope typically uses refracted light.
   ☐ True
   ☐ False

9. What is the difference between protozoa and metazoa?

10. Suctoria are good indicators of nitrification.
    ☐ True
    ☐ False

11. Metazoa found in the activated sludge process are believed to enter the system through infiltration of the collection system.
    ☐ True
    ☐ False

12. What is the difference between a free-swimming ciliate and a crawler ciliate?

13. Are rotifers protozoa or metazoa?

14. Protozoa can eat all the bacteria in an aeration basin.
   ☐ True
   ☐ False

15. A phase contrast microscope will provide more detail than a typical bright-field microscope.
   ☐ True
   ☐ False

16. Under proper environmental conditions, what physical characteristic allows bacteria to grow so quickly?

17. Metazoa found in the activated sludge process are believed to enter the system through infiltration of the collection system.
   ☐ True
   ☐ False

18. Lower life forms in activated sludge, such as individual bacteria cells, typically provide insight to the biomass age and stability.
   ☐ True
   ☐ False

19. The activated sludge process selects for organisms that swim well and consume the bacteria that create the sludge floc.
   ☐ True
   ☐ False

20. Nematodes can be found in the effluent of a WRRF due to their resistance to chlorination.
   ☐ True
   ☐ False

21. Annelids are segmented worms with soft, muscular body walls.
   ☐ True
   ☐ False

22. In high enough concentrations, bristle worms can cause red or pink colored foam conditions.
   ☐ True
   ☐ False

23. In activated sludge facilities, it has been reported that when ciliates are present, only 20% to 50% of the *E. coli* are removed.
   ☐ True
   ☐ False

24. A WRRF may never have a large diversity of protozoa or metazoa, yet it operates very well.
   ☐ True
   ☐ False

25. Aeolosoma organisms have been known to reproduce so rapidly they can change the color of the MLSS and cause rapid settling conditions.
   ☐ True
   ☐ False

Sequencing Batch Reactors

1. Sequencing batch reactor basins will typically complete one cycle per day.
   ☐ True
   ☐ False

2. Settling of the MLSS occurs in a separate basin outside of the SBR.
   ☐ True
   ☐ False

3. Nitrification is encouraged inside an SBR system during periods of increased mixed liquor temperatures.
   ☐ True
   ☐ False

4. Sequencing batch reactor basins are equipped with both mixing and aeration equipment that operate simultaneously.
   ☐ True
   ☐ False

5. In the conventional SBR design concept, there are three steps carried out during a batch.
   ☐ True
   ☐ False

6. Process upsets due to low pH in the SBR can occur if monitoring of process alkalinity is not performed.
   ☐ True
   ☐ False

7. During an aerated-fill cycle, both the influent and decant valve are in the open position.
   ☐ True
   ☐ False

8. During the anoxic phase of treatment, the DO should be monitored and should be always greater than 0.2 mg/L.
   ☐ True
   ☐ False

9. The idle and sludge wasting steps occur between the mixed-fill and aerated-fill phases.
   ☐ True
   ☐ False

10. It is recommended that a residual value of 50 mg/L of alkalinity be maintained in the decanted effluent.
    ☐ True
    ☐ False

11. The effluent flowrate from an SBR will be higher than the influent flowrate.
    ☐ True
    ☐ False

12. A sequence of anaerobic and anoxic phases followed by settling will ensure complete nitrification.
    ☐ True
    ☐ False

13. For SBR effluent flow equalization, the equalization basin is sized to hold a maximum of one-half the decantable volume.
    ☐ True
    ☐ False

14. For efficient alkalinity addition, sodium hydroxide is preferred over sodium bicarbonate or sodium carbonate.
    ☐ True
    ☐ False

15. For nitrogen removal, the fill and react phases include static fill, mixed fill, and mixed react.
    ☐ True
    ☐ False

16. To avoid overloading an individual SBR, the fill cycles should be arranged to spread the daily peak loading into multiple SBR basins.
    ☐ True
    ☐ False

Integrated Fixed-Film Activated Sludge and Moving Bed Biofilm Reactor Systems

1. Which of IFAS or MBBR is a stand-alone system with no RAS?

2. What control parameter is used for BOD and nitrification?

3. An MBBR can only be used for post-denitrification.
    - ☐ True
    - ☐ False

4. What is typical aerobic SRT range for IFAS systems?

5. The key to the IFAS process is the increased total biomass inventory that can be achieved by combining the attached biomass with the suspended MLSS.
    - ☐ True
    - ☐ False

6. Integrated fixed-film activated sludge systems are often implemented to improve nitrification capacity at existing facilities.
    - ☐ True
    - ☐ False

7. Fabric, web-type media used in IFAS systems is difficult to install but is not prone to fouling by rags.
    - ☐ True
    - ☐ False

8. Polypropylene finned cylindrical shaped media provides excellent mixing and can be converted to an MBBR system.
    - ☐ True
    - ☐ False

9. It can be difficult to maintain the aeration system when using free-floating media.
    - ☐ True
    - ☐ False

10. A difference between IFAS and MBBR systems is the lack of RAS recycle with MBBR systems.
    - ☐ True
    - ☐ False

11. Integrated fixed-film activated sludge media are commonly made of plastic, foam, or fabric material.
    - ☐ True
    - ☐ False

12. Polyvinyl sheet IFAS media have a high initial cost and are maintenance intensive.
    - ☐ True
    - ☐ False

13. In MBBR systems, biofilms primarily develop on the protected surface inside of the plastic biofilm carrier.
    - ☐ True
    - ☐ False

14. Integrated fixed-film activated sludge and MBBR treatment facilities cannot be incorporated to existing facilities and must be constructed as stand-alone unit processes.
    - ☐ True
    - ☐ False

15. In a denitrifying MBBR reactor, the biofilm media is uniformly distributed by mechanical mixers.
    - ☐ True
    - ☐ False

16. Screens with mesh sizes of 20 to 35 mm are used for retaining the plastic carriers within the MBBR basin.
    □ True
    □ False

17. Tertiary or add-on MBBR processes receiving wastewater that received significant upstream treatment do not require additional screening.
    □ True
    □ False

18. Because the specific gravity of MBBR media is between 0.94 and 0.96 g/cm³, both clean and biofilm-covered plastic biofilm carriers will sink in calm water.
    □ True
    □ False

19. Because of the turbulent motion of the carriers in the reactor, the biofilm thickness will become excessive.
    □ True
    □ False

20. Automated wiping systems are used to remove accumulated debris from cylindrical sieves.
    □ True
    □ False

21. Integrated fixed film activated sludge systems are limited in the amount of nitrification that can be achieved.
    □ True
    □ False

22. An MBBR system installed in Norway that had been routinely inspected showed excessive wear on the plastic biofilm carrier in less than 5 years of continuous operation.
    □ True
    □ False

23. Coarse-bubble diffusers used in MBBR systems are prone to scaling and fouling because of the large orifice dimension, and therefore require more maintenance than fine-bubble diffusers.
    □ True
    □ False

CHAPTER 5
Nontraditional Disinfection

Purpose and Function

1. Disinfection will destroy and inactivate which type of pollutants?
   a. Biochemical oxygen demand (BOD)
   b. Total suspended solids (TSS)
   c. Metals
   d. Bacteria and pathogens

2. Name a nontraditional disinfection technology.
   a. Chlorine gas
   b. Sodium hypochlorite
   c. UV
   d. Ozone

3. Which disinfection has become an accepted form of wastewater disinfection over the past several years?
   a. Chlorine gas
   b. Sodium hypochlorite
   c. UV
   d. PAA

Ozone Disinfection

1. Which of the following statements about ozone is correct?
   a. Ozone can remove contaminants of emerging concern.
   b. Ozone is practical if the facility's effluent contains high levels of suspended solids, chemical oxygen demand, or organic matter.
   c. Ozone is a less complicated disinfection method that chlorine and requires a more complex contact system.
   d. Costs for ozone are low because of the equipment and power requirements.

2. The type and quantity of the target pathogen are the only factors that influence the ability of ozone to disinfect.
   ☐ True
   ☐ False

3. What are the two common methods to introduce ozone in the wastewater for disinfection?
   a. Sidestream injection and mixers
   b. Sidestream injection and diffusers
   c. Mixers and diffusers
   d. Direct injection and mixers

Peracetic Acid Disinfection

1. The three types of mixing that can be considered for use in PAA contact basins are
   a. Hydraulic, static, and mechanical
   b. Hydraulic, diffused air, and mechanical
   c. Jet mixing, diffused air, and static
   d. Diffused air, mechanical, and jet mixing

Other Emerging Disinfectants

1. An advantage of UVC LED on wastewater disinfection is its
    a. Ability to be "plug and play" for some point-of-use applications
    b. Higher energy use
    c. Shorter lamp life
    d. Low cost

2. On-site generation of chlorine gas has gained acceptance due to
    a. Ease of use
    b. Supply chain issues resulting in short supply and higher costs
    c. Similar chemical distribution system
    d. Cost

3. Hydro-optic disinfection UV has been successfully used for
    a. Wastewater disinfection
    b. Combined sewer overflow disinfection
    c. Reuse disinfection
    d. Dechlorination

Combined Disinfection Methods and Advanced Oxidation

1. Combined disinfection for systems include
    a. PAA + chlorine
    b. Ozone + PAA
    c. Chlorine + UV
    d. PAA + PFA

2. Advanced oxidation processes are used to remove
    a. Trace organic pollutants
    b. Inorganics
    c. Heavy metals
    d. Chlorine residual

3. What would be a reason to install a combined disinfection system?
    a. Disinfection of fit-for-purpose water
    b. Economics
    c. Ease of operation
    d. Least amount of maintenance

CHAPTER 6
Membranes

Introduction, Purpose, and Function

1. A key advantage of membrane processes is that they do not generate a waste stream.
    - ☐ True
    - ☐ False

2. When a membrane system replaces the secondary clarifier in an activated sludge process, the resulting process is called
    - a. Microfiltration
    - b. Membrane bioreactor
    - c. Reverse osmosis
    - d. Wastewater recycling

3. Which of these types of membranes is best for removing disinfection byproducts?
    - a. Microfiltration
    - b. Ultrafiltration
    - c. Nanofiltration
    - d. Brackish water reverse osmosis

4. A membrane bioreactor can have a higher MLSS concentration, shorter HRT, and longer SRT than a conventional activated sludge system with a secondary clarifier.
    - ☐ True
    - ☐ False

5. What is membrane flux?
    - a. The holes in a membrane that allow water to pass through
    - b. A chemical to clean pipes before soldering
    - c. The amount of water that passes through a specified area of membrane per unit time
    - d. The membranes used in an activated sludge system

6. Which membrane has pores so small that only water molecules pass through?
    - a. Membrane bioreactor
    - b. Reverse osmosis
    - c. Microfiltration
    - d. Nanofiltration

7. What is potable reuse?
    - a. The intentional recycling of highly treated wastewater into a municipal water supply
    - b. The removal of pathogens from wastewater
    - c. The reuse of wastewater for irrigation of parks and green spaces
    - d. The reuse of wastewater for industrial purposes

Theory of Operation

1. Which one of the following statements is true?
    - a. Reverse osmosis membranes have larger pores the microfiltration membranes.
    - b. The mechanism for particle removal in membrane filtration is straining at the surface of the membrane.
    - c. Reverse osmosis membranes are able to achieve complete removal of all dissolved compounds.
    - d. Particle removal in membrane filtration is significantly affected by operating conditions like feed pressure and temperature.

2. Osmotic pressure helps water flow through a reverse osmosis system so that the water flow will be faster when the osmotic pressure is higher.
   ☐ True
   ☐ False

3. The cake layer that forms on the surface of a microfiltration or ultrafiltration membrane can be removed by backwashing.
   ☐ True
   ☐ False

4. Which type of reverse osmosis fouling is most responsible for limiting the water recovery in reverse osmosis systems?
   a. Accumulation of particulate matter
   b. Oxidation of reduced metals
   c. Scaling from precipitation of insoluble minerals
   d. Adsorption of organic and biological matter

5. Calculating the specific flux helps identify whether fouling is occurring.
   ☐ True
   ☐ False

6. Which type of reverse osmosis fouling is measured with the SDI?
   a. Accumulation of particulate matter
   b. Oxidation of reduced metals
   c. Scaling from precipitation of insoluble minerals
   d. Adsorption of organic and biological matter

7. Which of these salts or minerals can precipitate and cause scaling on reverse osmosis membranes?
   a. Sodium chloride
   b. Calcium phosphate
   c. Free chlorine
   d. Calcium tetrachloride

Expected Performance

1. The water recovery from membrane filtration systems is higher than reverse osmosis systems.
   ☐ True
   ☐ False

2. Reverse osmosis systems are backwashed more frequently than membrane filtration systems.
   ☐ True
   ☐ False

3. What can cause bacteria to get through a membrane?
   a. Operating at too low of a pressure
   b. Backwashing too frequently
   c. Not enough backwashing
   d. Tears or holes in the membrane

4. What chemical do manufacturers use to test the performance of reverse osmosis membranes?
   a. Sodium chloride
   b. Calcium phosphate
   c. Chlorine (bleach)
   d. Pharmaceuticals

Process Variables

1. What is the conductivity rejection for a reverse osmosis system if the feed conductivity is 1855 µS/cm and the permeate conductivity is 27.6 µS/cm?
    a. 96.5%
    b. 97.5%
    c. 98.5%
    d. 99.5%

2. In a challenge test for a membrane filtration system, the *Cryptosporidium* concentration is measured as 172 000 oocysts/L in the feed and 19 oocysts/L in the permeate. What LRV is achieved?
    a. 1.28
    b. 3.96
    c. 5.24
    d. 6.51

3. What is the water recovery of a reverse osmosis system if the feed flowrate is 475 gpm and the permeate flowrate is 360 gpm?
    a. 24%
    b. 32%
    c. 76%
    d. 132%

4. What is the water recovery of a reverse osmosis system if the feed flowrate is 1960 m³/d and the permeate flowrate is 2580 m³/d?
    a. 24%
    b. 32%
    c. 76%
    d. 132%

5. A reverse osmosis system has 24 pressure vessels; each pressure vessel has 7 membrane elements, and each membrane element has 37.1 m² of membrane surface area. What is the permeate flux if the permeate flowrate is 3250 m³/d?
    a. 0.0217 L/m²·h
    b. 21.7 L/m²·h
    c. 152 L/m²·h
    d. 521 L/m²·h

6. What is the flux for a membrane filtration skid if the skid produces 275 gpm and has 94 membrane elements that each have a membrane surface area of 66 sq ft?
    a. 46 gpd/sq ft
    b. 51 gpd/sq ft
    c. 61 gpd/sq ft
    d. 64 gpd/sq ft

7. What is the specific flux at 20 °C for a membrane filtration system that has a flux of 72.6 LMH when the water temperature is 13 °C and the TMP is 84 kPa (0.84 bar)?
    a. 70 LMH/bar
    b. 75 LMH/bar
    c. 89 LMH/bar
    d. 106 LMH/bar

8. What is the specific flux at 20 °C for a membrane filtration system that has a flux of 41 gpd/sq ft when the water temperature is 26 °C and the TMP is 11.5 psi?
    a. 3.0 gpd/sq ft/psi
    b. 3.6 gpd/sq ft/psi
    c. 4.3 gpd/sq ft/psi
    d. 34 gpd/sq ft/psi

9. What is the NDP of a reverse osmosis system that is operating with a feed pressure of 107 psi, permeate pressure of 9.1 psi, feed osmotic pressure of 11.3 psi, and permeate osmotic pressure of 0.81 psi?
    a. 85.8 psi
    b. 87.4 psi
    c. 97.1 psi
    d. 108 psi

Equipment for Membrane Filtration and Membrane Bioreactors

1. If removal of viruses is desired, microfiltration membranes should be used.
    ☐ True
    ☐ False

2. Isotropic membranes have the same mechanical properties throughout the membrane.
    ☐ True
    ☐ False

3. Which of the following is true for an outside-in membrane configuration?
    a. Air scour applies air so that it travels from the outside of the hollow fiber to the inside of the hollow fiber.
    b. The lumen is plugged with resin on both potted ends.
    c. Water is filtered from the outside of the hollow fiber to the inside and materials are rejected at the membrane surface.
    d. Filtrate is collected on the outside of the hollow fiber membranes.

4. Which of the following is true for a submerged system versus a pressurized system?
    a. Submerged systems have a narrower operational TMP range than pressurized systems.
    b. Submerged systems operate with a feed pump to draw water into the lumens.
    c. Both submerged and pressurized systems are dead-end filters.
    d. Submerged systems can come in both outside-in and inside-out configurations.

5. Membrane bioreactor systems use a drum screen upstream of biological reactors to remove larger particles.
    ☐ True
    ☐ False

6. Which component is not commonly found in membrane filtration systems?
    a. Strainer/drum screen
    b. Turbidimeter
    c. Compressed air system
    d. High-pressure feed pump

Process Control for Membrane Filtration and Membrane Bioreactors

1. What instruments are used to monitor the production cycle?
    a. Feed pressure transmitter, flow meter, and filtrate pressure transmitter
    b. Airflow rate, turbidimeter, and flow meter
    c. Feed and filtrate pressure transmitter
    d. Feed pump speed, filtrate valve percent opening, and turbidimeter

2. For a membrane system providing pathogen protection, if the filtrate turbidimeter is greater than 0.15 NTU for a period longer than 30 minutes, the membrane is considered intact.
    ☐ True
    ☐ False

3. Which of the following statements about the PDR is not true?
    a. The PDR is calculated by dividing the loss of pressure divided by the test period.
    b. If the PDR is greater than the UCL, the membranes are intact.
    c. Per U.S. EPA regulations, the PDR should be measured once a day to satisfy the direct monitoring requirement.
    d. The PDR can be used to evaluate if there is a potential breach in a membrane skid.

4. What parameter is used to evaluate the effectiveness of chemical cleans?
   a. TMP
   b. Turbidity
   c. Filtrate pressure
   d. Feed flow

Operation of Membrane Filtration and Membrane Bioreactors

1. What type of information can be determined with daily rounds?
   a. Type of foulants on the membrane
   b. Time until membrane failure
   c. All routine maintenance of analyzers have been completed
   d. Membrane integrity

2. Short-term storage of membranes can range from 1 day to 1 month.
   ☐ True
   ☐ False

3. Filtrate turbidity is the only parameter that can be used to evaluate the integrity of membranes.
   ☐ True
   ☐ False

Maintenance for Membrane Filtration and Membrane Bioreactors

1. Which of the following are examples of commonly used chemical cleaning methods for membrane filtration systems?
   a. Backwashing and air scour
   b. Enhanced flux maintenance and CIP
   c. Taking units out of service and letting them rest
   d. Reducing the flux at which membranes operate

2. What is the typical frequency for maintenance washes?
   a. Monthly to annually
   b. Daily to weekly
   c. Weekly to monthly
   d. Every backwash cycle

3. What chemicals are typically used for cleaning hollow fiber membranes?
   a. Caustic and sulfuric acid
   b. Sodium hypochlorite and citric acid
   c. Detergent solutions and hydrochloric acid
   d. Organic solvent and water

4. The process of repairing broken hollow fibers is called
   a. Pinching
   b. Parking
   c. Pinning
   d. Plucking

5. What procedures should be taken for membrane storage?
   a. Membranes should be drained of process water
   b. Membranes should be taken off and wrapped
   c. Membranes do not require special procedures for storage
   d. Membranes should be soaked in a mild chlorine solution or with a preservative such as sodium bisulfite

Troubleshooting for Membrane Filtration and Membrane Bioreactors

1. Which of the following water quality parameters should be checked when troubleshooting suspected fouling of the membrane systems?
   a. Feed ammonia, nitrite, and nitrate
   b. Feed TSS, TOC, and turbidity
   c. Iron, manganese, and pH
   d. Total coliforms, fecal coliforms, and chlorine residual

2. Which of the following water quality parameters should be checked when troubleshooting suspected scaling of the membrane systems?
   a. Iron and manganese and other metals
   b. TOC, ammonia, and TSS
   c. Total coliforms, fecal coliforms, and chlorine residual
   d. Temperature, pH, and electrical conductivity

3. What are the typical causes for drifting or abnormally high turbidity readings on the online turbidity meters?
   a. Seasonal changes in temperature, pH, and electrical conductivity
   b. Buildup of biofilms in the sample lines and the presence of air bubbles
   c. Loss of calibration
   d. Defective meter parts

4. What steps should be taken if a membrane filtration system fails a required daily pressure decay test in a potable reuse facility?
   a. Taken immediately out of service for troubleshooting of a failed PDT and repair of the cause
   b. Allowed to return to production and scheduled for maintenance at the next convenient time
   c. Stop production and order replacement of the membranes in the failed train
   d. Perform a backwash and EFM to restore membrane performance

Spiral-Wound Elements

1. Similar to low-pressure filtration membranes, reverse osmosis membranes can be backwashed.
   ☐ True
   ☐ False

2. How many membrane sheets must feedwater pass through to become permeate in a spiral-wound reverse osmosis element?
   a. Only one
   b. Two to four
   c. At least 10
   d. All of them

3. What are some common properties of polyamide-based spiral-wound reverse osmosis elements?
   a. Sensitive to free chlorine and ozone
   b. Very resistant to bacterial degradation
   c. Operationally stable when pH is above 11
   d. Consists of a single-layer membrane

4. What is the purpose of a feed spacer for a reverse osmosis element?
   a. Helps separate out large debris from entering the membrane
   b. Separates the membrane layers and creates a flow passage for feedwater
   c. Provides structure to the membrane that would otherwise collapse
   d. Spaces to space out feedwater from permeate

Pressure Vessels

1. Pressure vessels house multiple spiral-wound elements in series.
    - ☐ True
    - ☐ False

2. Which seals/O-rings prevent unintentional passage of feedwater into the permeate channel?
    a. Permeate interconnector O-ring
    b. Brine seal
    c. Endcap O-ring
    d. Permeate interconnector and endcap O-rings

3. What is the purpose of retaining rings in the context of reverse osmosis system pressure vessels?
    a. To provide a tight seal between spiral-wound elements
    b. To keep the endcaps from sliding outward
    c. To prevent backflow within the pressure vessel
    d. To connect the permeate tubes of adjacent pressure vessels

Stages and Trains

1. Which is the best array configuration to reach highest recovery?
    a. Two-pass system
    b. Single-stage system
    c. Multistage system
    d. Single pressure-vessel system

2. What is the vessel array configuration of a reverse osmosis train consisting of 100 vessels in Stage 1, 50 vessels in Stage 2, and 25 vessels in Stage 3?
    a. 4:2:1
    b. 1:2:4
    c. 100:50:25
    d. 25:50:100

3. What is the best configuration of pressure vessels for a three-stage system containing a total of 140 pressure vessels?
    a. 20:40:80
    b. 70:70
    c. 80:40:20
    d. 140:70:35

4. How is a reverse osmosis train defined?
    a. Having at least two or more vessels per stage
    b. A bank of pressure vessels that share common valving
    c. A railway wagon filled with reverse osmosis membrane elements
    d. A reverse osmosis system consisting of multiple stages

Energy Recovery Devices, Clean-in-Place Skid, Permeate Flush Skid

1. Energy recovery devices allow a reverse osmosis system to recoup part of the residual concentrate pressure.
    - ☐ True
    - ☐ False

2. What is the purpose of a CIP skid?
    a. To periodically clean the membranes by recirculating heated chemicals
    b. To further treat permeate that is not up to spec
    c. To clean the reverse osmosis skid after the membranes are removed
    d. To restore rejection of target compounds

3. Why is flushing with permeate important upon reverse osmosis system shutdowns?
   a. To increase recovery
   b. To avoid passive fouling and septicity
   c. To increase performance upon restart
   d. To reduce recovery

Operation of Reverse Osmosis

1. Reverse osmosis is typically operated at constant feed pressure and variable flux.
   ☐ True
   ☐ False

2. What type of activity is done during daily rounds for reverse osmosis systems?
   a. Initiate permeate flushes to restore performance
   b. Perform maintenance of online instrumentation
   c. Collect samples for compliance purposes
   d. Verify drawdowns of related chemicals, such as anti-scalant

3. What is a proper way to store reverse osmosis membranes?
   a. Follow storage protocols from membrane manufacturer
   b. Store in a chlorine solution of at least 1000 mg/L
   c. Store in an acid solution of pH below 2
   d. Store in a warehouse after membranes have fully dried in the field

4. At what frequency should reverse osmosis membranes be cleaned using CIP procedures?
   a. Frequency is site-specific, and may occur monthly to yearly
   b. Every time the reverse osmosis system is shut down for maintenance
   c. Every 24 hours to meet compliance requirements
   d. Reverse osmosis membranes should not be cleaned as cleaning may damage them

Maintenance for Reverse Osmosis

1. Reverse osmosis membranes can be cleaned with citric acid and sodium hypochlorite using similar protocols for microfiltration or ultrafiltration membranes.
   ☐ True
   ☐ False

2. What is the typical frequency for chemical cleaning of reverse osmosis membrane systems?
   a. Daily to weekly
   b. Weekly to monthly
   c. Every backwash cycle
   d. Monthly to annually

3. What method should be performed during prolonged downtime of the reverse osmosis membrane systems?
   a. Draining the reverse osmosis system
   b. Pickling with a biocide agent, such as SBS
   c. Soaking reverse osmosis system with chlorine solution
   d. Performing a CIP

4. What is the common cause of failure of O-ring seals in a reverse osmosis system?
   a. Extreme temperatures during operation
   b. Continuous exposure to combined chlorine
   c. Physical damage due to movement of water through the system
   d. Biological degradation

Troubleshooting for Reverse Osmosis

1. What are common causes for fouling of reverse osmosis membranes?
   a. Particulate and colloidal matter
   b. Pore plugging
   c. Growth of biological films and deposits of organics
   d. Accumulation of salt and clay deposits

2. What are common causes for scaling of reverse osmosis membranes?
   a. Loss of chemical feeds such as anti-scalant agent or sulfuric acid
   b. Too little of chloramines residual
   c. Too high of TDS
   d. Feed water pH too low

3. What methods can be used for troubleshooting reverse osmosis permeate conductivity increases?
   a. Measurement of conductivity in feed and permeate to calculate removal
   b. Measurement of conductivities from individual vessels and from vessel probing
   c. Check the SDI of the feed water
   d. Check chloramines residual and increase if too low

Safety Considerations

1. What are examples of typical CIP chemicals for membrane filtration systems?
   a. Sulfuric acid and caustic
   b. Caustic and sodium hypochlorite
   c. Sodium hypochlorite and citric acid
   d. Citric acid and caustic

2. What are examples of typical chemicals used for reverse osmosis systems to control scaling?
   a. Sulfuric acid and caustic
   b. Caustic and sodium hypochlorite
   c. Sulfuric acid and anti-scalant agent
   d. Citric acid and caustic

3. What documentation should always be reviewed before working with any chemicals?
   a. Chemical's certificate of analysis
   b. SDSs
   c. O&M manual
   d. Chemical compatibility guide

4. What is commonly done to prevent electrical hazards and injuries when servicing equipment?
   a. Turn off the equipment and shut off the breaker.
   b. Display a caution tape around working area.
   c. Deenergize at power source and lock out tag out.
   d. Consult SDSs.

5. What is commonly done before disposal of spent CIP solutions?
   a. Record temperature and conductivity before disposal.
   b. Neutralize by adjusting pH and adding dechlorinating agent.
   c. Dilute solution and send to drain.
   d. No neutralization is necessary.

CHAPTER 7
Water Reuse

Water Reuse Overview

1. Decentralized nonpotable reuse is most commonly referred to as "on-site reuse."
   - ☐ True
   - ☐ False

2. Which of the following describes "de facto" potable reuse?
   a. The planned use of recycled water for potable uses
   b. The planned use of recycled water for irrigation
   c. The use of a surface water that is affected by upstream wastewater discharge for potable supply
   d. The use of stormwater for a potable supply

3. The concept of "fit-for-purpose" treatment refers to
   a. Recycled water treatment to meet effluent guidelines defined by U.S. EPA
   b. Offsetting the use of potable water with advanced treated recycled water
   c. Treatment of recycled water to meet specific water quality objectives based on the intended uses of water
   d. Recycled water treatment for industrial reuse purposes

Secondary and Tertiary Treated Municipal Wastewater

1. Nitrified effluent contains free ammonia and typically has higher concentrations of TOC.
   - ☐ True
   - ☐ False

2. High concentrations of nutrients in secondary and tertiary effluents can lead to algal blooms.
   - ☐ True
   - ☐ False

Nonpotable Water Reuse

1. Historically, the largest use of potable water supply was for industrial uses.
   - ☐ True
   - ☐ False

2. Recycled water that comes in contact with the edible part of a crop will likely need to be disinfected before use.
   - ☐ True
   - ☐ False

3. Tertiary treatment may include
   a. Coarse bar screen
   b. Trickling filter
   c. Chlorine disinfection
   d. Grit removal

4. Recycled water has a higher potential for biogrowth during storage and distribution than typical potable supplies.
    ☐ True
    ☐ False

Potable Water Reuse

1. Direct injection is the most effective groundwater recharge method to create a barrier to prevent seawater intrusion to a confined aquifer.
    ☐ True
    ☐ False

2. The difference between indirect potable reuse and direct potable reuse is the use of an environmental buffer.
    ☐ True
    ☐ False

3. What are the "4Rs" in the potable reuse framework?
    a. Reliability, redundancy, recalcitrant, resilience
    b. Reliability, redundancy, robustness, resilience
    c. Responsibility, redundancy, robustness, resilience
    d. Reliability, reinvestigate, robustness, resilience

Groundwater Recharge

1. Algae can cause reduced infiltration rates in spreading basins.
    ☐ True
    ☐ False

2. Full advanced treatment consists of reverse osmosis and advanced oxidation.
    ☐ True
    ☐ False

3. When should direct injection be a considered method for groundwater recharge?
    a. The aquifer is unconfined
    b. The depth of the aquifer is less than 300 ft
    c. There is available land to perform surface spreading
    d. The depth of the aquifer is greater than 800 ft

4. Which is the most accurate method to estimate the travel time of reclaimed water recharged to the groundwater?
    a. Added tracer
    b. Intrinsic tracer
    c. Numerical modeling
    d. Darcy's law

Surface Water Augmentation

1. There are federal regulations established for planned surface water augmentation projects.
    ☐ True
    ☐ False

2. Posttreatment is an important consideration for surface water augmentation projects to ensure the water does not cause corrosion of conveyance pipelines or the drinking water distribution system.
    ☐ True
    ☐ False

3. Water should be dechlorinated before being released to a reservoir or river.
   ☐ True
   ☐ False

Direct Potable Reuse

1. There are established federal regulations for direct potable reuse.
   ☐ True
   ☐ False

2. Source control could be considered the first barrier in a direct potable reuse project and can provide protection by preventing unwanted contaminants from entering the collection system.
   ☐ True
   ☐ False

CHAPTER 8
Characterization and Sampling of Sludge

Types of Sludge

1. What is the typical removal efficiency for settleable solids in a primary clarifier?
    a. 10% to 15%
    b. 20% to 50%
    c. 40% to 60%
    d. 95% to 99%

2. Anaerobically digested sludge can be less _____ than aerobically digested sludge even it contains high ammonia and hydrogen sulfide.
    a. Odorous
    b. Viscous
    c. Decomposed
    d. Heated

3. Thermal hydrolysis of sludge is operated at a temperature above
    a. 50 °C
    b. 80 °C
    c. 120 °C
    d. 35 °C

4. Sludge concentration in a DAF unit typically ranges from
    a. 3% to 5%
    b. 5% to 8%
    c. 8% to 10%
    d. Above 12%

Sludge Characteristics

1. At what SVI value of sludge sample represents sludge bulking of an aeration tank?
    a. 100 mL/g
    b. 125 mL/g
    c. >150 mL/g
    d. 130 mL/g

2. At what temperature is a sludge sample burnt in a furnace to measure VSS?
    a. 550 °C
    b. 103 °C
    c. 250 °C
    d. 400 °C

3. Which parameter indicates a quick measurement of total oxygen demand required for an aeration tank?
   a. BOD
   b. COD
   c. SSOUR
   d. TTSS

4. Which parameter helps bacteria to bind together to create flocs?
   a. COD
   b. SVI
   c. pH
   d. EPS

5. Total suspended solids and total solids concentration of a sample can be found directly by weighing the sample before and after drying at
   a. 103 °C
   b. 153 °C
   c. 503 °C
   d. 450 °C

Sampling of Sludge

1. Which sampling method is typically suitable for sludge when the compositions do not vary over time?
   a. Composite
   b. Grab
   c. Diluted
   d. Monitoring

2. In the clarifier, the composite sampling is typically made for a period of
   a. 6 hours
   b. 12 hours
   c. 18 hours
   d. 24 hours

3. Sludge characterization is critical for following what treatment process?
   a. Overflow of primary clarifier
   b. Overflow of secondary clarifier
   c. Underflow of final clarifier
   d. Grit removal effluent

4. Sludge samples should be stored at a temperature of
   a. −4 °C
   b. 0 °C
   c. 4 °C
   d. 25 °C

CHAPTER 9
Management of Solids

Characterization

1. Which of the following is the ultimate goal of the U.S. EPA regulations related to biosolids?
   a. Minimize cost
   b. Speed of treatment
   c. Maximize the quantity treated
   d. Protect human health

2. In biosolids treatment, different classes reflect different levels of treatment. Which class represents the most stringent level?
   a. Class A
   b. Class B
   c. Class EQ
   d. Class Z

3. The primary federal regulation pertaining to biosolids is
   a. 40 CFR 503
   b. The Clean Water Act
   c. Vector Attraction Reduction Law
   d. National Pollutant Discharge Elimination System

4. How do state regulations relate to federal regulations?
   a. State regulations can be less stringent.
   b. State regulations can only be more stringent.
   c. The state cannot make its own regulations.
   d. States are always responsible for making separate regulations.

5. Which of the following is one of the pollutants regulated in biosolids under the Part 503 rule?
   a. Phosphorus
   b. Potassium
   c. Selenium
   d. Nitrogen

6. An active pretreatment program is implemented to
   a. Minimize unwanted constituents in the influent
   b. Reduce odors in the facility
   c. Remove pathogens
   d. Reduce the flowrate through the facility

7. Which of the following are essential to a successful land-application program?
   a. All testing on the biosolids must be done by outside laboratories.
   b. The biosolids must be in the form of a fine dust.
   c. The type of crops and nutrient needs of those crops should be considered.
   d. The site must have fencing of at least 10 ft in height.

8. Class A and B biosolids must meet which additional requirements besides the pathogen requirements?
   a. Surface water quality criteria
   b. Vector attraction reduction
   c. Crop application rate standards
   d. Carbon reduction criteria

Handling, Management, and Safety

1. Which of the following are required by federal regulations for compliance?
    a. A quality control and quality assurance plan
    b. Standard operating procedures for each analysis
    c. Chain-of-custody forms or a logbook
    d. An approved permit

2. Once a WRRF has met the general permit requirements, its personnel can initiate a land-application program by doing which of the following:
    a. Determine which farmers want the material. The site must have enough acres to justify the expenditure of time and resources that go into obtaining a permit.
    b. Have private meetings with only the farmer; the public does not need to know about the land application.
    c. Visit the land-application site only upon initiation of the project; there is no need for regular inspections.
    d. Purchase the crops for the proposed land-application site.

3. Which of the following is a beneficial use option for ash?
    a. Landfill cover
    b. Plastic food containers
    c. Tire additive
    d. Vitamin casing

4. The National Institute of Occupational Safety and Health recommends which of the following to provide a more comprehensive set of precautions for use by employers and employees?
    a. Employees purchase their own PPE
    b. Hygiene stations
    c. Liability waivers
    d. Optional training programs

5. What is the difference between land reclamation and land application?
    a. Biosolids application rates are higher
    b. Biosolids application rates are lower
    c. No mixing with amendments of any type
    d. Cannot use alkaline-stabilized biosolids

6. Which of the following product options is the most feasible when using thermal treatment or drying?
    a. Engineered soils
    b. Turf grass
    c. Aglime agent
    d. Biochar

7. Which environmental management system is the most appropriate for WRRFs to use?
    a. ISO 14001
    b. ISO 9000
    c. ISO 17021
    d. EMAS Article 4

8. Which type of PPE is appropriate for a job in which there is the potential for exposure to spray?
    a. Respirator
    b. Face shield
    c. High-visibility jacket
    d. Fall harness

CHAPTER 10
Additional Stabilization Methods

Composting

1. Before composting, sludge is typically dewatered to
    a. 5% to 10%
    b. 15% to 30%
    c. 75% to 90%
    d. 50% to 65%

2. When performing the squeeze ball test on compost to determine if sufficient moisture is present, you open your hand and the material in your hand retains the shape of a ball. The compost mixture has sufficient moisture.
    ☐ True
    ☐ False

3. The compost blend must have an appropriate ratio of carbon to nitrogen to encourage biological activity and the initial mix should be maintained with a C:N in the range of
    a. 5 to 10
    b. 20 to 30
    c. 10 to 40
    d. 50 to 50

4. The purpose of bulking agents is to provide porosity, structural support, and the source of carbon.
    ☐ True
    ☐ False

5. The operator should clean hoppers and conveyor(s) from debris and obstructions
    a. Weekly
    b. Monthly
    c. Annually
    d. As needed

6. A squeeze ball test is a quick measure of
    a. Temperature
    b. Moisture content
    c. Compost blend
    d. pH

7. If the compost pile is emitting odors, the operator should
    a. Decrease aeration frequency.
    b. Spray the pile with water or increase spray frequency.
    c. Increase airflow rate or run blowers continuously.
    d. Decrease cake dryness.

8. If anaerobic conditions occur during composting, the following will result:
    a. High compost temperature
    b. Large amounts of fine dust generated
    c. Generation of foul odors
    d. Reduced moisture content

9. The target moisture level of the compost blend is
   a. 45%
   b. 60%
   c. 35%
   d. 65%

Alkaline Stabilization

1. Lime addition to sludge reduces odors and pathogen levels by creating a high pH environment that
   a. Prevents chemical reaction from occurring
   b. Physically separates water from the sludge
   c. Prevents biological activity from taking place
   d. Promotes biological activity to take place

2. The reactivity level of lime is defined as high when a temperature rise of 40 °C (72 °F) is achieved in
   a. 30 minutes or less
   b. 3 minutes or less
   c. 60 minutes or less
   d. 6 minutes or less

3. The lime stabilization process is an exothermic reaction (produces heat).
   ☐ True
   ☐ False

4. If lime deposits in the lime slurry feeder, the operator should
   a. Reduce the slaker dilution water.
   b. Increase the velocity through the feeder by increasing the recirculation rate back to the to the slurry holding tank.
   c. Reduce the velocity through the feeder.
   d. Increase the dry lime feed rate to the lime slaker.

5. To avoid clogging of lime slurry transport lines, the system should have recirculation loops, flushing connections, avoid sharp bends, and use minimum 51-mm (2-in.) -diameter lines.
   ☐ True
   ☐ False

6. The alkaline stabilization process involves the addition of chemical to sludge to
   a. Reduce the pH below 7
   b. Increase the pH to above 12
   c. Neutralize the sludge
   d. Reduce the pH below 4

7. To meet Class B stabilization requirements, the pH of the feed dewatered sludge cake must be elevated to more than pH 12.0 for 2 hours and then maintained above pH 11.5 for
   a. 22 hours
   b. 24 hours
   c. 12 hours
   d. 16 hours

8. Equipment used in lime stabilization processes requires high maintenance because of the severe conditions caused by handling
   a. Acidic chemicals
   b. Caustic lime
   c. Hydrogen sulfide
   d. Explosive gases

9. Pre-lime stabilization and post-lime stabilization refer to where the lime is applied in the treatment process with respect to
   a. Dewatering
   b. Thickening

    c.   Secondary treatment

    d.   Anaerobic digestion

Heat Drying

1. During heat drying, two processes occur simultaneously, including the transfer of heat from the surrounding environment to evaporate the surface moisture and the transfer of internal moisture to the surface of the solid where it is evaporated.

    ☐ True

    ☐ False

2. To meet Class A standards with heat drying, the dried product must be 90% solids or greater and a temperature of

    a.   80 °C (176 °F) is exceeded

    b.   55 °C (131 °F) is maintained for 15 days or longer

    c.   180 °C (356 °F) or higher is achieved for 30 minutes

    d.   55 °C (131 °F) is maintained for 3 days or longer

3. Heat drying systems are designed to meet the product characteristics needed for the selected end use.

    ☐ True

    ☐ False

4. The following dryers are typically applied at smaller facilities:

    a.   Rotary drum dryer, fluidized bed dryer, and paddle dryer

    b.   Solar dryer, paddle dryer, and belt dryer

    c.   Solar dryer, fluidized bed dryer, and belt dryer

    d.   Rotary drum dryer, fluidized bed dryer, and belt dryer

5. Typical solids concentrations of 2% to 10% best describes what process?

    a.   Incineration

    b.   Dewatering

    c.   Thickening

    d.   Drying

6. Drying systems that heat the sludge by convection are referred to as

    a.   Indirect dryers

    b.   Solar dryers

    c.   Combination dryers

    d.   Direct dryers

7. The purpose of back-mixing is to

    a.   Increase the amount of exhaust gases for treatment

    b.   Increase the feed moisture content

    c.   Adjust the moisture content of the feed and promote stable pellet formation

    d.   Reduce the system's thermal efficiency

8. Which dryer has more complicated O&M than other dryers?

    a.   Solar dryer

    b.   Belt dryer

    c.   Paddle dryer

    d.   Rotary drum dryer

9. Belt dryers operate at gas temperatures of

    a.   93 °C (200 °F)

    b.   538 °C (1000 °F)

    c.   127 to 177 °C (260 to 350 °F)

    d.   871 °C (1600 °F)

Incineration

1. The typical excess air range in an FBI is
    a.   50% to 150%
    b.   0% to 100%
    c.   75% to 100%
    d.   40% to 50%

2. The MHI heat transfer is
    a.   High
    b.   Poor
    c.   Excellent
    d.   Average

3. The MHI combustion temperature range is between
    a.   730 to 760 °C (1350 to 1400 °F)
    b.   482 to 649 °C (900 to 1200 °F)
    c.   760 to 871 °C (1200 to 1400 °F)
    d.   760 to 871 °C (1400 to 1600 °F)

4. The FBI exit temperature range is between
    a.   816 to 900 °C (1500 to 1650 °F)
    b.   730 to 760 °C (1350 to 1400 °F)
    c.   482 to 649 °C (900 to 1200 °F)
    d.   127 to 177 °C (260 to 350 °F)

5. There typically is a temperature difference between the fluidized bed and the freeboard because a portion of VOCs will be combusted in the area above the bed, causing a temperature increase of 65 to 150 °C (150 to 300 °F).
    ☐   True
    ☐   False

6. The solids detention time in an FBI is
    a.   1 to 5 seconds
    b.   1 to 5 minutes
    c.   1 to 5 hours
    d.   40 to 60 minutes

7. The gas detention time in an MHI is
    a.   1 to 2 minutes
    b.   5 to 8 seconds
    c.   1 to 2 seconds
    d.   2 to 4 minutes

8. The typical excess air range in an MHI is
    a.   50% to 150%
    b.   0% to 100%
    c.   75% to 125%
    d.   40% to 50%

9. Refractory is a material that is resistant to heat. It is used as a lining in furnaces to resist high heat temperature (above 538 °C [1000 °F]), thermal shocks, erosion, physical attack, and chemical attack.
    ☐   True
    ☐   False

10. Incineration of 25% dewatered sludge cake will reduce the volume by
    a.   2 times
    b.   13 times
    c.   4 times
    d.   3 times

CHAPTER 11
Odor Control

Purpose and Function

1. Odor concentration, intensity, persistence, and character descriptors are measurable objective parameters of perceived odor.
    - ☐ True
    - ☐ False

2. The only enforceable H_2S air quality standards currently in effect are the OSHA occupational regulations.
    - ☐ True
    - ☐ False

3. Hydrogen sulfide intoxication is a concern in
    a. Open spaces
    b. Dark areas
    c. Cities
    d. Confined spaces

4. Hydrogen sulfide is both an irritant and a chemical asphyxiant, with effects on both oxygen use and the central nervous system.
    - ☐ True
    - ☐ False

5. At concentrations around _____, H_2S numbs the olfactory system and prevents odor detection.
    a. 10 ppmv
    b. 50 ppmv
    c. 100 ppmv
    d. 200 ppmv

6. Odor threshold of a compound is
    a. The highest concentration at which that compound can be detected by human smell
    b. The lowest concentration at which that compound causes severe headaches
    c. The lowest concentration at which that compound causes shortness of breath
    d. The lowest concentration at which that compound can be detected by human smell

7. Hydrogen sulfide is a predominant odor-causing constituent emitted at WRRFs.
    - ☐ True
    - ☐ False

8. _____ have the potential to generate and release odors to the surrounding area.
    a. Places or processes in which wastewater is collected, conveyed, or treated
    b. Offices
    c. School buildings
    d. Gardens

9. Hydrogen sulfide is _____ than air and, therefore, often collects in low-lying areas and working sites near the ground.
    a. Heavier
    b. Lighter
    c. Darker
    d. More volatile

10. To help minimize odor production, minimum full-pipe velocity should greater than _____ to prevent deposition of solids and the formation of H_2S.
   a. 3 m/s (10 ft/sec)
   b. 0.6 m/s (2 ft/sec)
   c. 1.8 m/s (6 ft/sec)
   d. 0.2 m/s (0.6 ft/sec)

11. Adaptation or olfactory fatigue is a phenomenon that occurs when people with a normal sense of smell experience
   a. An aromatic odor
   b. A citrus odor
   c. A rotten vegetable odor
   d. A decrease in perceived intensity of an odor

12. Odors from environmental sources can cause health symptoms depending on
   a. Source of odor
   b. Unpleasantness of odor
   c. Familiarity of odor
   d. Age, sex, medical condition, and the level and type of the odorous substance in the environment

Odor Control Collection Equipment

1. Odor collection systems should maintain negative pressure to _____.
   a. Prevent any air from entering the system
   b. Prevent fugitive emissions
   c. Create a vacuum chamber
   d. Stop odors from forming

2. An odor collection system should prevent air from entering or leaving the system.
   ☐ True
   ☐ False

Odor Treatment Technologies

1. Activated carbon removes odorous chemicals through the following surface-based process:
   a. Diffusion
   b. Adsorption
   c. Oxidization
   d. Absorption

2. Virgin carbon media are activated carbon media that have been chemically treated with additional chemical compounds.
   ☐ True
   ☐ False

3. Biofilters are very effective at removing sulfur-based odor compounds such as hydrogen sulfide, organic sulfides, and mercaptans, but generally are not effective at removing nitrogen-based compounds such as ammonia and amines.
   ☐ True
   ☐ False

4. Where is liquid-phase treatment typically used for odor control?
   a. Water resource recovery facility
   b. Wastewater collections system
   c. Emission stacks
   d. Anaerobic digesters

CHAPTER 12

Instrumentation

Lifecycle of an Instrument

1. The labor and resources required to operate and maintain an instrument can exceed the original purchase price and installation cost.
   - ☐ True
   - ☐ False

2. Transmitters are always installed in the same case (integral) with their sensors.
   - ☐ True
   - ☐ False

3. This is an example of a physical property.
   - a. pH
   - b. Pressure
   - c. Ammonia
   - d. Nitrate

4. This term is used to describe a measuring device that takes information from a process and converts it into a standardized signal.
   - a. Instrument
   - b. Sensor
   - c. Solenoid
   - d. Probe

5. Instruments that provide information to automated control systems all consist of at least these two components:
   - a. Sensor and controller
   - b. Controller and transmitter
   - c. Transducer and controller
   - d. Sensing element and transmitter

6. Optical sensors convert light into a/an
   - a. Pressure signal
   - b. Chemical signal
   - c. Electrical signal
   - d. Wavelength signal

7. Passive instruments differ from active instruments in this way.
   - a. Active instruments use pressure to generate power.
   - b. Passive instruments do not require a power source.
   - c. Active instruments only display the current condition.
   - d. Passive instruments send signals to controllers.

8. When using an analog signal, this reading indicates the zero value or the lowest possible measurement.
   - a. 0 mA
   - b. 4 mA
   - c. 16 mA
   - d. 20 mA

9. When an instrument reads consistently high or consistently low, the resulting data have
   a. Accuracy
   b. Precision
   c. Bias
   d. Damping

10. The lowest concentration that can be reliably measured as different from zero is the
    a. Detection limit
    b. Quantitation limit
    c. Accuracy
    d. Precision

11. A pressure instrument is capable of reading values between 70 and 1720 kPa (10 and 250 psi). What is the span?
    a. 70 kPa (10 psi)
    b. 1650 kPa (240 psi)
    c. 1720 kPa (250 psi)
    d. 1790 kPa (260 psi)

12. _____ can be included in a transmitter's electronics to remove signal noise and amplify an instrument's output.
    a. Bias cancellation
    b. Horns and antennae
    c. Sensors and probes
    d. Signal conditioners

13. _____ is the ability to set a lower limit to a positive number that is created by instrument installation position, for example, when an instrument is located below the low level of a tank.
    a. Span adjustment
    b. Instrument accuracy
    c. Zero suppression
    d. Signal damping

Location/Position and Speed/Movement

1. A _____ is an electromechanical device made up of an actuator that is mechanically linked to an electrical switch.
   a. Proximity switch
   b. Limit switch
   c. Level switch
   d. Flow switch

2. A _____ detects the presence of an object without actually making contact with it.
   a. Proximity switch
   b. Limit switch
   c. Level switch
   d. Flow switch

3. _____ is the term used when a switch is activated when it was not intended to be.
   a. Bias adjustment
   b. Zero adjustment
   c. Signal damping
   d. Nuisance tripping

4. The "normal" status of a switch is where it would be positioned
   a. After activation
   b. Most of the time
   c. With no power connected
   d. Under typical process conditions

5. In a "motion sensor" configuration, a proximity switch will sense the presence (and, indirectly, the speed) of metallic tabs connected to the _____ of a piece of rotating equipment.
   a. Starter
   b. Shaft
   c. Breaker
   d. Magnet

Pressure

1. _____ pressure is a pressure that is referenced to atmospheric pressure.
   a. Absolute
   b. Differential
   c. Gauge
   d. Switch

2. _____ is a force distributed over an area.
   a. Proximity
   b. Speed
   c. Flow
   d. Pressure

3. The pressure exerted by a column of liquid is calculated by multiplying the gravitational acceleration times the height of the liquid column times the _____.
   a. Dielectric constant
   b. Density of the liquid
   c. Electrical current
   d. Capacitive potential

4. Gas inside a sealed container will exert pressure
   a. Varying by location in container
   b. The same in all directions
   c. The same as a liquid
   d. Varying by density gradient

5. At sea level, atmospheric pressure would be equal to
   a. 0 kPaG (0 psig)
   b. 0 psia
   c. 0 atm
   d. Perfect vacuum

6. Pressure _____ are flat ceramic or corrugated metal disks that deflect when subjected to changing pressures.
   a. Switches
   b. Transmitters
   c. Diaphragms
   d. Seals

7. _____ on gauge pressure sensors must be kept open and dry so that atmospheric pressure can reach the reference side of the sensor.
   a. Vent tubes
   b. Electrical switches
   c. Actuators
   d. Transmitters

8. Which fluid is used in fluid-filled pressure instruments that operate in extreme cold conditions?
    a. Glycerin
    b. Salt water
    c. Halocarbon
    d. Silicone

9. The best thing to clean pressure sensors with is
    a. A dishwashing scrub pad
    b. Steel wool
    c. Soapy water
    d. A bottle brush

10. _____ occurs in pressure instruments when an instrument reads different values for the same pressure condition when measuring increasing pressures versus when reading decreasing pressures.
    a. Span adjustment
    b. Hysteresis
    c. Temperature correction
    d. Zero suppression

11. Block and bleed valves are used when a pressure instrument is
    a. Isolated and drained
    b. Used with corrosive chemicals
    c. Subjected to pressure spikes
    d. Disassembled and rebuilt

Temperature

1. The most accurate type of temperature sensor is
    a. Thermistor
    b. Thermowell
    c. Resistance temperature detector
    d. Thermocouple

2. _____ is the temperature sensor that is the most rugged (can handle more vibration and minor impacts).
    a. Thermistor
    b. Thermowell
    c. Resistance temperature detector
    d. Thermocouple

3. Thermowells can be filled with _____ to increase the sensor's response time.
    a. Salt water
    b. Conductive paste
    c. Electrolyte solution
    d. Spray foam

4. Resistance temperature detectors are often made out of _____ because this provides the broadest temperature range and the best repeatability.
    a. Gold
    b. Platinum
    c. Copper
    d. Nickel

Level Instruments

1. A _____ is activated when water level rise causes it to move from a vertical position to a horizontal position.
   a. Pressure transmitter
   b. Flow switch
   c. Float switch
   d. Conductance transmitter

2. _____ are often added to float switch installations to reduce the effect of turbulence in the wet well/tank and to help ensure the switch is consistently activated at the same elevation.
   a. Buoyancy correction factors
   b. Cable weights
   c. Automatic flushing devices
   d. Cable retractors

3. _____ can prevent a float from being activated at the proper liquid level.
   a. pH fluctuations
   b. Temperature extremes
   c. Inconsistent liquid column density
   d. Floating debris and stringy material

4. _____ both provide point level control.
   a. Ultrasonic level instruments and float switches
   b. Noncontact radar and pressure-sensing level transmitters
   c. Conductivity and float switches
   d. Conductivity and impedance level transmitters

5. Conductivity switches provide a noncontact level sensing.
   ☐ True
   ☐ False

6. Bubbler systems are a new technology that is becoming popular for continuous level measurement.
   ☐ True
   ☐ False

7. _____ uses high frequency sounds waves and "time of flight" to calculate a tank liquid level.
   a. Ultrasonic level sensor
   b. Noncontact radar level sensor
   c. Conductivity switch
   d. Proximity switch

8. The _____ on a "time-of-flight" sensor is created by the fact that the signal cannot be sent and received at the same time.
   a. Bias
   b. Zero suppression
   c. Deadband
   d. Hysteresis

9. The sensor on a "time-of-flight" instrument should not be installed in the middle of a domed cover because that placement can create _____.
   a. Instrument grounding issues
   b. Secondary echo issues
   c. Interference with cover supports
   d. Magnetic field generation

10. Radar and ultrasonic sensors must send signals perpendicular to the liquid surface.
   ☐ True
   ☐ False

11. _____ on a noncontact radar instrument increase(s) the range of the instrument.
  a. Higher voltage power
  b. Heavier cable
  c. Larger antennae
  d. Stronger magnets

12. When setting up the 4 to 20 mA output from a radar level sensor, the 50% tank level should have what mA output signal?
  a. 4 mA
  b. 8 mA
  c. 12 mA
  d. 16 mA

13. A tank full of clean water has an open top. A pressure-sensing level instrument is used to measure the water level. If the water level in the tank is 1.22 m (4 ft), what pressure will the transmitter be reading?
  a. 11.93 kPa (1.73 psi)
  b. 14.82 kPa (2.15 psi)
  c. 19.34 kPa (2.81 psi)
  d. 23.48 kPa (3.41 psi)

14. It is necessary to use a differential pressure transmitter when you are measuring level in
  a. A tank that is open to atmospheric pressure
  b. A tank that is covered and vented
  c. A tank that is covered and pressurized
  d. A tank that is open and in a cold climate

15. To measure continuous tank level, the submersible pressure transducer must be installed
  a. Below the lowest tank level
  b. At the midpoint of the tank depth
  c. At the maximum tank level
  d. In a thermowell next to the tank

16. Pressure-sensing systems in wastewater systems with diaphragm seals and sealed sensing fluid lines benefit from protecting sensing lines against blockages and preventing possible corrosion.
  ☐ True
  ☐ False

Flow Meters

1. For proper operation, most closed-pipe flow meters required _____ flow through/past the flow element.
  a. Turbulent
  b. Laminar
  c. High-velocity
  d. Low-velocity

2. Where installation conditions do not allow for the minimum straight run of pipe requirements for a closed-pipe flow meter to be met, a _____ can be installed to improve the flow characteristics.
  a. Flow conditioner
  b. Flow classifier
  c. Velocity exciter
  d. Venturi port

3. In an electromagnetic flow meter, the liquid flow through the pipe acts as a(n) _____.
  a. Permanent magnet
  b. Grounding rod
  c. Electrical capacitor
  d. Electrical conductor

4. In an electromagnetic flow meter, the _____ of the liquid moving through the pipe is calculated from the measured induced voltage, the flow tube diameter, and the magnetic field strength.
   a. Temperature
   b. pH
   c. Velocity
   d. Turbulence

5. In an electromagnetic flow meter, the induced voltage is measured across the _____.
   a. Junctions
   b. Terminals
   c. Diaphragms
   d. Electrodes

6. Magmeters are capable of measuring reverse flows.
   ☐ True
   ☐ False

7. The flow element of a thermal mass flow meter contains a heated sensor RTD and a _____.
   a. Gas temperature sensing RTD
   b. Magnetic coil
   c. Electrode
   d. Density compensating element

8. _____ should be installed for critical flow meters that are installed on pipes that cannot be taken out of service.
   a. Isolation valves
   b. Quick disconnect couplings
   c. Backup generators
   d. Bypass lines

9. A _____ is a passive testing device that is set up with flexibility to select combinations of resistors to create required testing values.
   a. Block and bleed valves
   b. Resistance decade box
   c. Handheld calibration unit
   d. Magnetic resonance indicator

10. Thermal mass flow meters are more effective at measuring flowrates in gases than in liquids.
    ☐ True
    ☐ False

11. The flow element for a thermal mass flow meter must be installed in a specific orientation to the flow direction for the meter to operate properly.
    ☐ True
    ☐ False

Dissolved Oxygen Meters

1. For electrochemical DO probes to work properly, the liquid being tested must flow over the membrane continuously.
   ☐ True
   ☐ False

2. Dissolved oxygen passes easily through membranes coated with grease.
   ☐ True
   ☐ False

3. With optical DO sensors, light output increases with increasing DO concentration.
   ☐ True
   ☐ False

4. Electrochemical DO sensor membranes must be kept wet to prevent loss of electrolyte through the membrane.
   - ☐ True
   - ☐ False

5. When water quality parameters are measured, this comes into direct contact with the water
   a. Instrument
   b. Processor
   c. Relay
   d. Probe

6. Dissolved oxygen sensors read
   a. Percent saturation
   b. Oxygen concentration
   c. Mass of oxygen
   d. Barometric pressure

7. The two types of DO sensors used in wastewater treatment include
   a. Clark and galvanic
   b. Polarographic and galvanic
   c. Optical and electrochemical
   d. Optical and polarographic

8. Which of the following parameters has the greatest effect on DO concentration?
   a. Salinity
   b. Barometric pressure
   c. Temperature
   d. pH

9. Warm water can hold _____ DO than cold water at the same pressure and salinity.
   a. More
   b. Less
   c. Half
   d. Double

10. As barometric pressure increases, the DO concentration at saturation _____.
    a. Increases
    b. Decreases
    c. Remains unchanged
    d. Matches salinity

11. Samples that contain the maximum amount of oxygen for a given temperature, pressure, and salinity are said to be _____ with DO.
    a. Undersaturated
    b. Saturated
    c. Supersaturated
    d. Hyposaturated

12. The purpose of the electrolyte in polarographic and galvanic sensors is to
    a. Equalize the salt concentration in the sample
    b. Transfer corrosion byproducts away from the cathode
    c. Form a connection between the anode and cathode
    d. Provide a pathway for oxygen to move up the probe

13. This type of DO probe requires a 5- to 15-minute warmup period before use.
    a. Galvanic
    b. Optical
    c. Reeding
    d. Polarographic

14. How long must electrochemical DO probes be conditioned before they are calibrated for the first time?
   a. 6 hours
   b. 12 hours
   c. 18 hours
   d. 24 hours

15. This critical parameter must be entered by the operator when calibrating DO probes.
   a. pH
   b. Temperature
   c. Percent salinity
   d. Barometric pressure

16. This substance may accumulate on the anode of a polarographic (Clark) electrode over time.
   a. Silver chloride
   b. Biofilm
   c. Gold bromide
   d. Salt

17. The sensor cap for an optical DO sensor should be replaced
   a. Every 30 days
   b. Every 3 months
   c. Once a year
   d. Every 2 years

18. Aggressive cleaning of optical DO sensors can
   a. Damage the DO permeable membrane
   b. Remove luminescent material from the cap
   c. Improve sensitivity at low DO concentrations
   d. Deposit silver chloride on the gold anode

Chlorine Residual Analyzers

1. This variable has the biggest effect on the percentages of hypochlorous acid and hypochlorite ion in water.
   a. pH
   b. Temperature
   c. DO
   d. Alkalinity

2. This form of chlorine dominates at low pH.
   a. Chloramine
   b. Hypochlorite ion
   c. Hypochlorous acid
   d. Free

3. Which of the following can be measured?
   a. Dose
   b. Demand
   c. Residual
   d. Extra

4. N,N-diethyl-p-phenylenediamine reagent turns _____ in the presence of residual chlorine.
   a. Yellow
   b. Blue
   c. Orange
   d. Magenta

5. At high concentrations of residual chlorine, the DPD test may appear this color
    a. Magenta
    b. Clear
    c. Orange
    d. Blue

6. One difference between the free chlorine and total chlorine test methods is
    a. The free chlorine test measures chloramines.
    b. The total chlorine test has a shorter react period.
    c. The free chlorine test turns magenta, then clear.
    d. The total chlorine test reagent includes potassium iodide.

7. Amperometric chlorine analyzers measure _____ as it passes through the membrane and reacts with the anode.
    a. Free chlorine
    b. Total chlorine
    c. Hypochlorous acid
    d. Hypochlorite ion

8. Amperometric chlorine residual measurements are sensitive to _____.
    a. Temperature
    b. Alkalinity
    c. Dissolved oxygen
    d. Ammonia

9. Small-diameter tubing used for reagents must be cleaned or replaced regularly to prevent
    a. Reagent contamination
    b. Biofilm growth
    c. Quenching
    d. Bypassing

pH Sensors

1. A sample at pH 4 contains _____ times more hydrogen ions than a sample at pH 5.
    a. 1
    b. 10
    c. 100
    d. 1000

2. The bulb of a pH sensor is made from
    a. Plastic
    b. Teflon
    c. Glass
    d. Metal

3. pH measurements use a reference electrode to measure the _____ between the electrolyte and the liquid being tested.
    a. Gravimetric potential
    b. Chloride disparity
    c. Piezometric charge
    d. Potential difference

4. The hydrogen ion concentration is higher on the outside of a glass pH sensor than it is on the inside. The solution being measured is
    a. Acidic
    b. Basic
    c. Neutral
    d. Charged

5. pH sensors must be calibrated periodically using
   a. Process water
   b. Buffer solutions
   c. Zero oxygen solution
   d. Water-saturated air

6. The main component of an ISFET pH sensor is the
   a. Gel layer
   b. Glass bulb
   c. Transistor
   d. Electrolyte

7. This substance can interfere with pH measurements by preventing hydrogen ions from interacting with the sensor.
   a. Alkalinity
   b. Suspended solids
   c. Grease
   d. Detergent

8. Which of the following cleaning methods should be used to remove mineral deposits from a pH sensor?
   a. Dilute hydrochloric acid
   b. Rubbing alcohol
   c. Mild dish detergent
   d. Soft polymer brush

9. Ion-selective field effect transistor sensors require a _____-point calibration.
   a. One
   b. Two
   c. Three
   d. Four

10. pH measurements may be affected by _____.
    a. Flow
    b. Suspended solids
    c. Dissolved oxygen
    d. Temperature

Oxidation–Reduction Potential Analyzers

1. Two ORP probes made from the same materials and with the same fill solution will always measure the same ORP value.
   ☐ True
   ☐ False

2. Oxidation–reduction potential probes are similar to pH probes with an electrode and reference electrode.
   ☐ True
   ☐ False

3. Both ORP and pH must be corrected for temperature effects and incorporate temperature sensors.
   ☐ True
   ☐ False

4. When a substance is oxidized it
   a. Gains oxygen
   b. Loses oxygen
   c. Gains water
   d. Loses water

5. What will the ORP most likely be in an aeration basin with a DO concentration of 2 mg/L?
   a. −300 mV
   b. −150 mV
   c. 0 mV
   d. +150 mV

6. Maintenance of ORP sensors requires
   a. Regular calibration
   b. Polishing
   c. Acid soaking
   d. Electrolysis

Total Suspended Solids Meters

1. Optical solids sensors must be recalibrated when solids characteristics change.
   ☐ True
   ☐ False

2. Microwaves travel faster through water than sludge.
   ☐ True
   ☐ False

3. This parameter can affect solids measurements by optical sensors.
   a. Color
   b. Conductivity
   c. Air bubbles
   d. Temperature

4. Transmittance-based solids analyzers can measure solids concentrations up to
   a. 1%
   b. 5%
   c. 10%
   d. 15%

5. This type of solids analyzer is capable of accurately reading solids concentrations as low as 0.5 mg/L.
   a. Transmittance
   b. Reflectance
   c. Microwave
   d. Electromagnetic

6. Microwave solids analyzers are sensitive to
   a. Upstream fittings
   b. Dissolved oxygen
   c. Conductivity changes
   d. System pressure

7. Optical solids analyzers should be calibrated at least
   a. Weekly
   b. Monthly
   c. Seasonally
   d. Annually

8. Which of the following solids analyzers should be selected for monitoring dewatered cake?
   a. Transmittance
   b. Reflectance
   c. Optical
   d. Microwave

Interface/Sludge Blanket Level Analyzers

1. Optical interface sludge blanket analyzers are lowered into the clarifier blanket each time a measurement is desired.
    - ☐ True
    - ☐ False

2. Rags, plastic, and grease can interfere with ISBL measurements.
    - ☐ True
    - ☐ False

3. Clarifier blanket depths may be manually measured with a
    - a. Rake arm
    - b. Sludge judge
    - c. Interface/sludge blanket level analyzer
    - d. Radar

4. Ultrasonic interface sludge blanket analyzers use _____ to determine sludge blanket depth in clarifiers and gravity settlers.
    - a. Sound waves
    - b. Radar waves
    - c. Light waves
    - d. Proton waves

Gas Detectors

1. _____ at WRRFs monitor for hydrogen sulfide, methane, and low oxygen conditions.
    - a. Differential pressure instruments
    - b. Dissolved oxygen analyzers
    - c. Fuel emissions testing
    - d. Gas detectors

2. _____ are areas that are not designed for regular occupancy, are large enough for a person to enter, and have limited means for entry and exit.
    - a. Confined space areas
    - b. Office space cubicles
    - c. Hazardous materials storage areas
    - d. Control rooms

3. _____ gas sensors are located in the area that they are testing, and gases enter the sensors by naturally moving from areas of higher concentration to areas of lower concentration.
    - a. Forced air
    - b. Pumped in
    - c. Diffusion mode
    - d. Capillary action

4. _____ sensors are basically fuel cells where the sensing electrode is treated with a catalyst that will react with a specific target gas.
    - a. Catalytic
    - b. Electrochemical
    - c. Infrared gas
    - d. Optical scatter

5. _____ sensors have an active bead that is treated with a catalyst that will oxidize when exposed to a flammable/combustible gas.
    - a. Catalytic
    - b. Electrochemical
    - c. Infrared gas
    - d. Optical scatter

6. _____ sensors monitor for changes to the infrared light beam that occur when target combustible gases/ hydrocarbons are present.
   a. Catalytic
   b. Electrochemical
   c. Infrared gas
   d. Optical scatter

7. A _____ is a passive device with many serpentine pathways that break a flame up and cool it down to extinguish it.
   a. Combustion chiller
   b. Flame alarm
   c. Fire extinguisher
   d. Flame arrester

8. Sensor _____ is immediate and irreversible damage that occurs when a catalytic sensor is exposed to chemicals that block the pores of the bead or bind to the catalyst and reduce the activity of the sensor.
   a. Poisoning
   b. Sickness
   c. Death
   d. Destruction

9. Under the right conditions, gas sensors could last forever.
   ☐ True
   ☐ False

10. For most sensors, they start to be consumed from the first day they are produced.
    ☐ True
    ☐ False

11. A _____ is used to verify that a gas sensor is operating correctly.
    a. Gas conductance test
    b. Bump test
    c. Resistive decade box
    d. Magnetic reactivity monitor

CHAPTER 13

Supervisory Control and Data Acquisition (SCADA) Systems

Purpose and Function and Theory of Operation

1. Label the sections on the following high-level overview of a SCADA system.

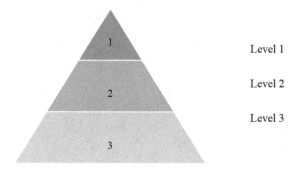

Level 1

Level 2

Level 3

2. Correctly label the five core components of a SCADA system as shown in the following image.

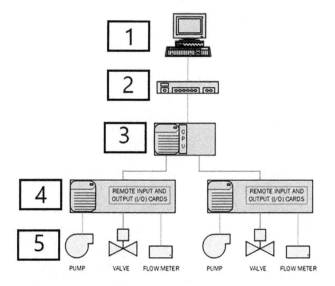

Equipment

1. Supervisory control and data acquisition equipment is the same as office computer systems.
    - ☐ True
    - ☐ False

2. Supervisory control and data acquisition systems are cyber-physical systems.
    - ☐ True
    - ☐ False

3. Control panels often include circuit breakers, fuses, and power supplies.
   - ☐ True
   - ☐ False

4. It is OK to connect any laptop computer to a SCADA Ethernet switch.
   - ☐ True
   - ☐ False

Operation

1. Moving between displays of the same level is called?
   a. Drill down
   b. Lateral navigation
   c. Starboard navigation
   d. Clickthrough

2. What is the key difference between a faceplate and a popup?
   a. Faceplates represent controls for a specific device
   b. Faceplates show video of the operator's face
   c. Faceplates only show controls
   d. Popups can't have controls for individual devices

3. What is the concept of "drill down" when referring to navigation?

4. What is lateral navigation and why is it important?

Troubleshooting

1. When inspecting a panel, externally, which of the following would cause the panel to fail:
   a. The panel exterior is dry, with no liquids dripping on it.
   b. There is a loud grinding sound coming from the fan.
   c. The panel-mounted display is in good working order.
   d. All of the panel latches are properly closed and secured.

CHAPTER 14
Leadership and Management

Permits and Reporting

1. Maximum Achievable Control Technology 129 is required by the NPDES permit for stormwater monitoring.
    ☐ True
    ☐ False

2. A residential discharger requires an industrial pretreatment permit.
    ☐ True
    ☐ False

3. A DMR report is required monthly only for WRRFs with permit violations.
    ☐ True
    ☐ False

4. Local limits are limits set by the utility or WRRF for industrial dischargers.
    ☐ True
    ☐ False

5. The Safe Drinking Water Act regulates point source pollution.
    ☐ True
    ☐ False

6. An operator should read and know the requirements of the entire NPDES permit, not just the effluent limits table.
    ☐ True
    ☐ False

7. A DMR contains important information related to the WRRF. Which of the following is found in the DMR?
    a. The permit number
    b. Unit process information
    c. Population served
    d. Facility size

8. In terms of biosolids permitting, WRRFs are required to report on a _____ basis.
    a. Monthly
    b. Quarterly
    c. Bi-monthly
    d. Annual

Sampling and Records

1. When setting up a sampling schedule, where is the first place to start?
    a. Outside laboratory
    b. Facility management
    c. NPDES permit or regulatory requirements
    d. Internet search

2. What is the most important information to know about an outside laboratory?

   a. Price

   b. QA/QC

   c. Method

   d. Shipping method

3. The pH, DO, and chlorine residual are generally tested on-site and not sent to an outside laboratory.

   ☐ True

   ☐ False

Budgeting and Financial Management

1. How does a capital budget differ from an operational budget?

   a. It is funded solely from user rates.

   b. Only a manager can develop a capital budget.

   c. It typically has a longer time horizon (5 to 10 years).

   d. It is always higher than an operating budget.

2. Operating budgets are predominantly funded by user rates.

   ☐ True

   ☐ False

3. Which of the following does not typically follow a procurement process?

   a. Purchase of chemicals

   b. Hiring staff

   c. Hiring a contractor

   d. Disposal of biosolids

4. A line-item budget is best defined by

   a. Its total cost line-by-line

   b. Being structured around specific activities at the utility

   c. Key performance indicators and performance standards

   d. Basic operational needs and revenues of the utility by individual line item

Writing and Updating Standard Operating Procedures

1. Who is the intended audience of an SOP?

   a. New employees

   b. Experienced employees

   c. Everyone in an organization

   d. People within an organization who will be performing the task described in the SOP

Asset Management and Maintenance

1. Preventive maintenance happens after equipment breaks.

   ☐ True

   ☐ False

2. Corrective maintenance happens after the equipment breaks.

   ☐ True

   ☐ False

3. By using a risk matrix, WRRFs can prioritize capital spending based on equipment criticality.

   ☐ True

   ☐ False

4. All equipment should be used until it dies. Replacing equipment before if fails is a waste of money.
    ☐ True
    ☐ False

Capital Improvements and Projects

1. Which of the following do not describe a type of procurement?
    a. Evaluating multiple bids based on predetermined criteria
    b. Reusing old materials on-site
    c. Pre-purchase of goods or services
    d. Contract allowance

2. Procurement of engineering services is typically not based on
    a. The firm's size and ability to perform the work
    b. The lowest bid
    c. The firm's experience on similar projects
    d. Scoring criteria developed by the utility

3. Which of the following is the most common delivery method for a project?
    a. Multi-prime
    b. Design-build-operate
    c. Design-bid-build
    d. Design-build

4. What is not a key element of an engineering contract?
    a. Project scope
    b. The experience of an engineer
    c. Schedule

5. Operations and maintenance staff have valuable insight and data to contribute to the design of a CIP project.
    ☐ True
    ☐ False

Emergency Planning and Response

1. Which of the following are not one of the phases of emergency management?
    a. Mitigate
    b. Recover
    c. Expect
    d. Prepare

2. How often should you update your ERP?
    a. Every 5 years
    b. When you have the time
    c. Never
    d. Yearly

3. What is not the responsibility of the incident commander?
    a. Make sure staff know the set proprieties
    b. Approving press releases
    c. Ensure everything is done in a safe manner
    d. Sharing information with the media

4. What is a natural disaster you might face at a WRRF?

    a. Active shooter

    b. Union strike

    c. Hurricane

    d. Cyberattack

5. A vulnerability assessment should consider only threats from human action, such as cyberattacks or bombs, because natural disasters cannot be prevented.

    ☐ True

    ☐ False

6. National Incident Management System and ICS are for police, fire fighters, and first responders only and are not suitable for a water utility.

    ☐ True

    ☐ False

7. One of the most important things learned from a vulnerability assessment is

    a. Learning how to manage peak flows

    b. Knowing where your facility is at risk

    c. Predicting what event will happen at your facility

    d. To help set guidelines about employee behavior

8. Match the following terms:

    a. Prevention

    b. Protection

    c. Mitigation

    d. Response

    e. Recovery

    1. To restore, strengthen, and revitalize infrastructure, housing, and a sustainable economy in a timely manner, as well as the health, social, cultural, historic, and environmental fabric of communities affected by a catastrophic incident.

    2. To avoid or stop an imminent, threatened, or actual act of terrorism.

    3. To reduce the loss of life and property by lessening the effect of future disasters.

    4. To keep safe our citizens, residents, visitors, and assets against the greatest threats and hazards in a manner that allows our interests, aspirations, and way of life to thrive.

    5. To react quickly to save lives, protect property and the environment, and meet basic human needs in the aftermath of a catastrophic incident.

9. Which member of the ICS should be commenting to the media?

    a. The incident commander

    b. Employees leaving the scene

    c. The public information officer

    d. The board chairperson

10. Your vulnerability assessment should be

    a. Shared with other facilities so you can see if you have common risk

    b. Read by every employee as part of their training

    c. Read by key officials and kept in a safe spot with limited access

    d. Made a public document on your website

Personnel Management

1. When planning your workforce, it is good to know the following:

    a. The experience level and service time of your staff

    b. If you have enough friends on you staff

    c. Who seems like trouble

    d. Do I have enough staff until I retire

2. When recruiting for your staff, it is good practice to
   a. Make sure my friends have a job.
   b. Know the social and economic makeup of the local community.
   c. Make sure I will have the same interest as the people I am recruiting.
   d. Recruit people who are the opposite of me so they challenge my decisions.

3. When disciplining employees, it is good to
   a. Be fair and firm.
   b. Punish people based on how my day is going.
   c. Make sure people who like me do not get in trouble.
   d. If I do not like someone, now is the time to take it out on them.

4. When dealing with a conflict between one of your employees and someone from another department, it is good to
   a. Yell more at the employee that is not in your department.
   b. Call the manager of the employee in the other department so they can deal with their employee.
   c. Because the employees are from different departments, it is okay for them to have conflict.
   d. Discipline both employees yourself.

5. When promoting an employee, it is best to
   a. Choose your friend.
   b. Choose the person that has been there the longest.
   c. Choose the candidate that is most qualified for the job.
   d. Promote everyone that applied so you do not have to make a tough decision.

6. When receiving feedback on a task from an employee, you should
   a. Consider the motivation behind the feedback.
   b. Disregard it; the employee needs to follow training and SOPs.
   c. Listen and work with the employee to see if their feedback helps.
   d. Tell the employee to do what they see is fit.

7. When working with unions, it is important not to
   a. Understand the employees' rights.
   b. Interpret the union contract to fit your perspective.
   c. Ask the human resources department for help.
   d. Follow proper procedure.

Leadership

1. In terms of WRRF leadership, what does "EQ" stand for?
   a. Equalization tank
   b. Emotional intelligence
   c. Intelligent quotient
   d. Equilibrium

2. When planning for your organization, you should set a short-term goal of (X) year(s) and a long-term goal of (X) year(s).
   a. 1, 10
   b. 5, 3
   c. 1, 3
   d. 5, 10

3. Which of the following is not a WRRF leader EQ trait?
   a. Self-regulation
   b. Motivation
   c. Empathy
   d. Strictness

4. What are the goals of a WRRF training program?
   a. To gain credit hours
   b. To get a free day off
   c. To learn skills to help both with leadership and job performance
   d. To advance toward a college degree

5. As a WRRF leader, you are not in charge of what?
   a. Facility design
   b. Safety
   c. Promoting
   d. Discipline

Customer Service and Public Engagement

1. Odor complaints should be ignored if the WRRF was built before the complainer's house was built.
   ☐ True
   ☐ False

2. Which parts of the facility should be part of a tour with middle school kids?
   a. Electrical room
   b. Chemical storage areas
   c. Basins without handrails
   d. Clarifiers with handrails

3. When giving a facility tour, an operator should use big words and complex explanations of the treatment process.
   ☐ True
   ☐ False

4. If a ratepayer approaches you in the grocery store and complains about their bill, what should you do?
   a. Politely refer them to customer service.
   b. Explain your utility's budgeting process and how you estimate power and chemicals for the fiscal year.
   c. Ignore them.
   d. Make a speech about regulatory burdens, workforce needs, and critical infrastructure.

List of Acronyms

% = percent

$(NH_4)_2SO_4$ = ammonium sulfate

°C degrees = Celsius

°F degrees = Fahrenheit

A/V = area-to-liquid volume ratio

ABC = Associated Boards of Certification (now Water Professionals International, or WPI)

ac = acre

ACH = air changes per hour

AERMOD = steady-state air dispersion model used to assess air pollutant effects

ANSI = American National Standards Institute

AOP = advanced oxidation process

APC = air pollution control

API = American Petroleum Institute

BACT = best available control technology

BMP = best management practice

BNR = biological nutrient removal

BOD = biochemical oxygen demand

BOD_5 = 5-day biochemical oxygen demand

BTFs = biotrickling filters

Btu = British thermal unit

BWRO = brackish water reverse osmosis

C = carbon

$C_6H_8O_7$ = citric acid

Ca^{2+} = calcium

$CaCO_3$ = calcium carbonate

cal/gmol = calorie/gram-mole

cCOD = colloidal chemical oxygen demand

CDC = Centers for Disease Control

CEB = chemically enhanced backwash

CECs = compounds of emerging concern

CEMS = continuous emissions monitoring system

CFR = Code of Federal Regulations

CFU = colony-forming unit

CH_3COOH = acetic acid

CH_4 = methane

CIP = capital improvement plan and clean-in-place

CIU = categorical industrial user

Cl_2 = chlorine gas

cm = centimeter

CMMS = computerized maintenance management system

CO = carbon monoxide

CO_3 = carbonate

COD = chemical oxygen demand

COS = carbonyl sulfide

CPI = corrugated plate interceptor

CPVC = chlorinated polyvinyl chloride

CS_2 = carbon disulfide

CT = concentration and time

cu ft = cubic feet

cu ft/h = cubic feet per hour

CWA = Clean Water Act

d = day

D/T = dilution-to-threshold ratio

DAF = dissolved air flotation

DALYs = disability adjusted life years

DCS = distributed control system

DCU = distributed control unit

deg = degree

DMR = discharge monitoring report

DO = dissolved oxygen

DPC = distributed process controller

DPD = N,N-diethyl-p-phenylenediamine

EBPR = enhanced biological phosphorus removal

EDC = endocrine-disrupting compound

EFM = enhanced flux maintenance

EMS = environmental management systems

EOC = emergency operations center

EPS = extracellular polymeric substance

ERD = energy recovery device

ERG = enforcement response guide

ERP = emergency response plan and Enforcement Response Plan

F:M = food-to-microorganism ratio

FC = fecal coliforms

Fe = iron

FEMA = Federal Emergency Management Agency

FOG = fats, oils, and grease

FRP = fiber-glass-reinforced plastic

ft = feet

ft/s = feet per second

G = Camp-Stein velocity gradient

g = gram

GAC = granular activated carbon

gal = gallon

gfd = gallons per square foot per day

GIS = geographic information system

GLUMRB = Great Lakes Upper Mississippi River Board

GOX = gaseous oxygen

gpd/sq ft = gallons per square foot per day

gpm = gallons per minute

h = hour

H^+ = hydrogen ion

H_2S = hydrogen sulfide

H_2SO_4 = sulfuric acid

HAZCOM = hazard communication program

HAZMAT = hazardous materials

HAZWOPER = hazardous waste operations and emergency response

HCl = hydrochloric acid
HDPE = high-density polyethylene
HMI = human–machine interface
HOCl = hypochlorous acid
HOD UV = hydro-optic disinfection ultraviolet light
h = hour
HRT = hydraulic retention time
HS^- = hydrosulfide
HVAC = heating, ventilating, and air conditioning
I&I = infiltration and inflow
IAP = incident action plan
IC = internal circulation
ICS = incident command structure
IFAS = integrated fixed-film activated sludge
in. = inch
ISBL = interface/sludge blanket level
ISFET = ion-selective field effect transistor
ISO = International Organization for Standardization
IU = industrial user
JHA = job hazard analysis
K = selectivity coefficient
kg = kilogram
kgal = kilogallon
KI = potassium iodide
kPa = kilopascals
kPaG = kilopascals gauge
kW = kilowatt
L = liter
L:W = length-to-width ratio
lb = pound
LED = light emitting diode
LEL = lower explosive limit
LIMS = laboratory information management system
LMH = liters per square meter per hour
LOI = local operator interface
LOTO = lockout/tagout
LOX = liquid oxygen
LRV = log removal value
m = meter
m/s = meters per second
m/s^2 = meters per second squared
m^2 = square meters
m^3 = cubic meter
m^3/d = cubic meters per day
m^3/h = cubic meter per hour
mA = milliamp
MACT = maximum achievable control technology
MAHL = maximum allowable headworks loading
MAIL = maximum allowable industrial loading
MBBR = moving bed biofilm bioreactor
MBR = membrane bioreactor
MCLs = maximum contaminant levels
MCRT = mean cell residence time
MEWP = mobile elevated work platform
mg = milligram
mg/L = milligrams per liter

Mg^{2+} = magnesium
mgd = million gallons per day
MHI = multiple-hearth incinerator
min = minute
mL = milliliter
ML/d = megaliters per day
MLR = mixed liquor recycle
MLSS = mixed liquor suspended solids
MLVSS = mixed liquor volatile suspended solids
mm = millimeter
mol = mole
MOPO = maintenance of plant operations
MPN = most probable number
MSDS = material safety data sheet
mV = millivolt
N = nitrogen
Na^+ = sodium
$NaHSO_3$ = sodium bisulfite
NaOCl = sodium hypochlorite
NaOH = sodium hydroxide
NBRC = National Blue Ribbon Commission for On-Site
        Non-Potable Water Systems
NDP = net driving pressure
NFPA = National Fire Protection Association
NH_3 = ammonia
NH_3-N = ammonia-nitrogen
NH_4^+ = ammonium ion
NH_4Cl = ammonium chloride
NH_4-N = ammonium as nitrogen
NH_4OH = ammonium hydroxide
NIMS = National Incident Management System
NIOSH = National Institute for Occupational Safety and Health
NLs = notification levels
NO_2^- = nitrite ion
NO_2-N = nitrite as nitrogen
NO_3^- = nitrate ion
NO_3-N = nitrate as nitrogen
NOM = natural organic matter
NPDES = National Pollutant Discharge Elimination System
NSPS = New Source Performance Standards
NTU = nephelometric turbidity unit
O&M = operations and maintenance
O_2 = oxygen
OCl^- = hypochlorite ion
OH^- = hydroxide ion
ONWS = on-site non-potable water reuse system
ORP = oxidation–reduction potential
OSHA = U.S. Occupational Safety and Health Administration
OWS = oil-water separator
P = phosphorus
Pa = pascal
PAA = peracetic acid
PAC = powder activated carbon
PCBs = polychlorinated biphenyls
pCOD = particulate chemical oxygen demand
PCS = process control system

PDR = pressure decay rate

PDT = pressure decay test

PEL = permissible exposure limit

PES = polyethersulfone

PFA = performic acid

PFAS = per- and poly-fluoroalkyl substances

PFOA = perfluorooctanoic acid

PFOS = perfluorosulfanoic acid

PFRP = process to further reduce pathogens

PID = process and instrumentation diagram

PIV = powered industrial vehicle

PLC = programmable logic controller

PO_4 = orthophosphate

PO_4^{-3} = phosphate ion

PO_4-P = orthophosphate as phosphorus

POTW = publicly owned treatment works

ppbv = parts per billion by volume

PPCPs = pharmaceuticals and personal care products

PPE = personal protective equipment

ppm = parts per million

ppmv = parts per million by volume

ppt = part per trillion

ppt = parts per thousand

PRCS = permit-required confined space

PRV = pressure-reducing valve

psi = pounds per square inch

psig = pounds per square inch gauge

PSRP = process to significantly reduce pathogens

PTFE = polytetrafluoroethylene

PVC = polyvinyl chloride

PVDF = polyvinylidene fluoride

Q = flow

QA = quality assurance

QA/QC = quality assurance/quality control

QACs = quaternary ammonium compounds

QAPP = quality assurance project plans

QC = quality control

RAS = return activated sludge

RFP = request for proposal

RFQ = request for qualifications

RTD = resistance temperature detector

RTU = remote telemetry unit

s = second

S^{-2} = sulfide

SAP = sampling and analysis plans

SBR = sequencing batch reactor

SBS = sodium bisulfite

SCADA = supervisory control and data acquisition

scfm = standard cubic feet per minute

sCOD = soluble chemical oxygen demand

SDI = silt density index

SDS = safety data sheet

SFPUC = San Francisco Public Utilities Commission

SIU = significant industrial user

SMP = soluble microbial products

SO_2 = sulfur dioxide

SOP = standard operating procedure

SOUR = specific oxygen uptake rate

sq ft = square feet

SRT = solids retention time

SSI = sewage sludge indicator

SUR = surface underflow rate

SVI = sludge volume index

SVOC = semivolatile organic compound

SWPPP = stormwater pollution prevention plan

SWRO = seawater reverse osmosis

TCLP = toxicity characteristic leaching procedure

TDS = total dissolved solids

THC = total hydrocarbon

TIR = total internal reflection

TKN = total Kjeldahl nitrogen

TMDLs = total maximum daily loads

TMP = transmembrane pressure

TOC = total organic carbon

TPHs = total petroleum hydrocarbons

TRS = total reduced sulfur

TSS = total suspended solids

TWL = top water level

U.S. EPA = United States Environmental Protection Agency

U.S. = United States

UASB = upflow anaerobic sludge blanket

UCL = upper control limit

UCMR = Unregulated Contaminant Monitoring Rule

UIC = underground injection control

UV = ultraviolet

UVC LED = ultraviolet C light-emitting diode

UVT = ultraviolet transmittance

V = volt

V/G/C = viruses, *Giardia*, and *Cryptosporidium*

VAR = vector attraction reduction

VDC = volts direct current

VFA = volatile fatty acid

VFD = variable-frequency drive

VOC = volatile organic compound

VSR = volatile solids reduction

VSS = volatile suspended solids

W:D = width-to-depth ratio

WAS = waste activated sludge

WET = whole effluent toxicity

WEV = water expulsion

WLAs = waste load allocations

WRRF = water resource recovery facility

Answer Keys

CHAPTER 1

1. False. The Pretreatment Act does not distinguish between commercial and industrial users. It only defines domestic and industrial wastewater.
2. B
3. True
4. True
5. False. Pretreatment programs, including permitting and administrative and enforcement tasks, can also be administered by U.S. EPA and NPDES state authorities.
6. B
7. False. All WRRFs treating more than 19 000 m³/d (5 mgd) or smaller WRRFs with significant industrial discharges must have pretreatment programs. Smaller WRRFs without significant industrial discharges are not required to.
8. True
9. A
10. D
11. True
12. True
13. True
14. False. Flow proportional sampling should be implemented wherever possible, except for those parameters that cannot be stored for the sampling duration and must be collected as grab samples. Also, to reflect the average water quality of the total volume of water discharged over the sampling period.
15. True
16. True
17. Water resource recovery facilities must inspect each industrial discharger at least once per year, per 40 CFR 403.8(f)(2)(v). Although U.S. EPA regulations establish a minimum frequency for inspections, additional inspections by the WRRF might be necessary depending on issues such as the variability of an SIU's effluent, the effect of the SIU's discharge on the WRRFs, and the facility's compliance history.
18. False. Regulation of PFAS continues to evolve as toxicity and treatability information develop.
19. False. Industrial facilities can implement their own wastewater treatment and discharge under an NPDES permit provided by a state or the U.S. EPA.
20. True
21. False. Treatment for these contaminants is most often through oxidation or sorption but depends on the specific chemistry of each contaminant.
22. False. Activated sludge treatment removes many degradable constituents associated with industrial wastewater and can, to some degree, absorb toxicity or polymer effects. However, polymers used at industrial facilities often pass through the treatment system and adversely affect polymer use at the WRRF for settleability, phosphorus precipitation, or solids dewatering processes.
23. False. Some food processors have high phosphorus and nitrogen compared to BOD and TSS. This is true in wastewaters high in protein, for example. Other food processes, such as sugar beet processing, result in wastewater with very high COD but very low nitrogen and phosphorus.
24. C
25. True
26. False. The effect of polymer flocculants on WRRF processes is most often interference with flocculants or coagulants used at the WRRF to enhance settling in secondary clarifiers or chemical phosphorus removal processes.
27. True
28. True
29. False. Equalization at an industrial site refers to using a large tank to store industrial wastewater before discharge to balance variability in discharge flow and water quality and minimize effects on downstream wastewater treatment operations.
30. True
31. B
32. True
33. True

34. False. "High-strength" refers to high amounts of organics, frequently reflected by high concentrations of COD and/or BOD.
35. False. Heavy metal wastewater should be source separated as much as possible for separate treatment. Source pretreatment processes like chemical precipitation, ion exchange, and membrane separation are more effective with concentrated heavy metal wastewaters.
36. True
37. D
38. C
39. False. Chemical precipitation is usually the first step considered for metals pretreatment. Ion exchange is considered a secondary polishing step for metals removal due to its higher cost compared to chemical precipitation.

CHAPTER 2

SAFETY PROGRAMS/INCIDENT PREVENTION

1. False. The main goal is to prevent workers from being injured.
2. True
3. A
4. C

HIERARCHY OF CONTROLS

1. True
2. B
3. True

PERSONAL PROTECTIVE EQUIPMENT

1. D
2. D

FALL PROTECTION AND PREVENTION

1. True
2. D
3. True
4. C
5. True
6. C

LOCKOUT/TAGOUT (CONTROL OF HAZARDOUS ENERGY)

1. B
2. B
3. False. Lockout/tagout is a legal requirement per the OSHA standard for The Control of Hazardous Energy (Lockout/Tagout), Title 29 Code of Federal Regulations (CFR) Part 1910.147 and addresses the practices and procedures necessary to disable machinery or equipment, thereby preventing the release of hazardous energy while employees perform servicing and maintenance activities.

CONFINED SPACE ENTRY

1. True
2. True
3. True
4. C
5. B

HAZARD COMMUNICATION

1. C
2. A
3. C

CHEMICAL HANDLING, STORAGE, AND RESPONSE

1. A
2. C
3. B
4. B

HEAVY EQUIPMENT AND MACHINERY

1. False. Equipment and machines do not always come from the manufacturer with the proper guards. These may need to be purchased in addition, or fabricated.
2. C
3. True
4. C
5. True
6. True
7. False. Cranes and hoists are required to be inspected frequently and periodically. Various items will be inspected anywhere from daily to monthly to annually based on OSHA and the manufacturer's instructions.
8. Label the parts of the PIV with the following terms:
   Fork   Mast   Overhead Guard   Counterweight   Tilt Cylinder   Load Backrest

9. False. Modified cranes can only be used after the modifications and associated structure are thoroughly checked for the new rated load by a qualified engineer or the equipment manufacturer.
10. False. Powered industrial vehicle training can be a combination of lecture, discussion, interactive computer learning, video, written material, and practical training. An evaluation must also be performed to evaluate the operator's competency to operate.
11. False. Operators/occupants should never increase their height by standing on rails or using a ladder or other means of increasing height. A larger or different type of MEWP should be used instead.

CHAPTER 3

CHEMICAL DOSING

1. A
2. A
3. A
4. D
5. D
6. D
7. C
8. D
9. C

COAGULATION AND FLOCCULATION

1. D
2. False

3. C
4. A
5. A
6. C
7. C
8. C
9. C
10. B

CLARIFICATION

1. C
2. A
3. C
4. A
5. False. Short-circuiting occurs when water does not mix properly, creating dead zones.
6. True
7. A
8. D
9. B
10. C

FILTRATION

1. A
2. D
3. B
4. D
5. A
6. Filter ripening, effective filtration, breakthrough
7. D
8. D
9. D

GRANULAR ACTIVATED CARBON

1. A
2. C
3. A
4. C
5. A
6. B
7. A
8. A
9. A
10. B

ION EXCHANGE

1. B
2. A
3. B
4. C
5. B
6. B
7. A
8. A
9. A
10. B

CHAPTER 4

1. A
2. True
3. False
4. False
5. False
6. Floc, filaments, higher life forms, and bulk water
7. False
8. False
9. Protozoa are single celled and metazoa are multicellular.
10. True
11. True
12. Free swimmers have cilia covering the entire organism, whereas the crawler ciliate has fused-together cilia called cirri.
13. Metazoa
14. False
15. True
16. Very large surface-to-volume ratio
17. True
18. False
19. False
20. True
21. True
22. True
23. False
24. True
25. True

1. False
2. False
3. True
4. False
5. False
6. True
7. False
8. False
9. False
10. True
11. True
12. False
13. False
14. False
15. True
16. True

1. MBBR
2. Dissolved oxygen
3. True
4. 2 to 5 days
5. True
6. True
7. False

8. True
9. True
10. True
11. True
12. False
13. True
14. False
15. True
16. False
17. True
18. False
19. False
20. False
21. False
22. False
23. False

CHAPTER 5

PURPOSE AND FUNCTION

1. D
2. D
3. D

OZONE DISINFECTION

1. A
2. False. The factors that affect the ability of ozone to disinfect include target pathogen, the pathogen's exposure to ozone, particulate shielding, and temperature.
3. B

PERACETIC ACID DISINFECTION

1. A

OTHER EMERGING DISINFECTANTS

1. A
2. B
3. D

COMBINED DISINFECTION METHODS AND ADVANCED OXIDATION

1. C
2. A
3. A

CHAPTER 6

INTRODUCTION, PURPOSE, AND FUNCTION

1. False
2. B
3. D
4. True
5. C
6. B
7. A

THEORY OF OPERATION
1. B
2. False
3. True
4. C
5. True
6. A
7. B

EXPECTED PERFORMANCE
1. True
2. False
3. D
4. A

PROCESS VARIABLES
1. C
2. B
3. C
4. C
5. B
6. D
7. D
8. A
9. B

EQUIPMENT FOR MEMBRANE FILTRATION AND MEMBRANE BIOREACTORS
1. False
2. True
3. C
4. A
5. True
6. D

PROCESS CONTROL FOR MEMBRANE FILTRATION AND MEMBRANE BIOREACTORS
1. A
2. False
3. B
4. A

OPERATION OF MEMBRANE FILTRATION AND MEMBRANE BIOREACTORS
1. D
2. False
3. False

MAINTENANCE FOR MEMBRANE FILTRATION AND MEMBRANE BIOREACTORS
1. B
2. B
3. B
4. C
5. D

TROUBLESHOOTING FOR MEMBRANE FILTRATION AND MEMBRANE BIOREACTORS
1. B
2. A
3. B
4. A

SPIRAL-WOUND ELEMENTS

1. False
2. A
3. A
4. B

PRESSURE VESSELS

1. True
2. D
3. B

STAGES AND TRAINS

1. C
2. C
3. C
4. B

ENERGY RECOVERY DEVICES, CLEAN-IN-PLACE SKID, AND PERMEATE FLUSH SKID

1. True
2. A
3. B

OPERATION OF REVERSE OSMOSIS

1. False
2. D
3. A
4. A

MAINTENANCE FOR REVERSE OSMOSIS

1. False. Polyamide reverse osmosis membranes may not be cleaned with sodium hypochlorite due to chemical damage to the membranes.
2. D
3. B
4. B

TROUBLESHOOTING FOR REVERSE OSMOSIS

1. C
2. A
3. B

SAFETY CONSIDERATIONS

1. C
2. C
3. B
4. C
5. B

CHAPTER 7

WATER REUSE OVERVIEW

1. True
2. C
3. C

SECONDARY AND TERTIARY TREATED MUNICIPAL WASTEWATER
1. False
2. True

NONPOTABLE WATER REUSE
1. False
2. True
3. C
4. True

POTABLE WATER REUSE
1. True
2. True
3. B

GROUNDWATER RECHARGE
1. True
2. True
3. B
4. A

SURFACE WATER AUGMENTATION
1. False
2. True
3. True

DIRECT POTABLE REUSE
1. False
2. True

CHAPTER 8

TYPES OF SLUDGE
1. D
2. A
3. C
4. A

SLUDGE CHARACTERISTICS
1. C
2. A
3. C
4. D
5. A

SAMPLING OF SLUDGE
1. B
2. D
3. C
4. C

CHAPTER 9

CHARACTERIZATION

1. D
2. C
3. A
4. B
5. C
6. A
7. C
8. B

HANDLING, MANAGEMENT, AND SAFETY

1. D
2. A
3. B
4. B
5. A
6. B
7. A
8. B

CHAPTER 10

COMPOSTING

1. B
2. True
3. B
4. True
5. D
6. B
7. C
8. C
9. A

ALKALINE STABILIZATION

1. C
2. B
3. True
4. B
5. True
6. B
7. A
8. B
9. A

HEAT DRYING

1. True
2. A
3. True
4. B
5. C
6. D
7. C

8. D
9. C

INCINERATION

1. D
2. B
3. D
4. A
5. True
6. B
7. C
8. C
9. True
10. B

CHAPTER 11

PURPOSE AND FUNCTION

1. True
2. True
3. D
4. True
5. C
6. D
7. True
8. A
9. A
10. B
11. D
12. D

ODOR CONTROL COLLECTION EQUIPMENT

1. B
2. False. A collection system needs to allow air to enter the system to replace the odorous air that is removed by the conveyance system.

ODOR TREATMENT TECHNOLOGIES

1. B
2. False. Virgin carbon media are activated carbon media that have not been used for treatment or regenerated previously.
3. True
4. B

CHAPTER 12

LIFECYCLE OF AN INSTRUMENT

1. True
2. False. They may also be wall-mounted nearby. Additional remote indicators may be installed in other areas.
3. B
4. A
5. D
6. C
7. B
8. B
9. C
10. A

11. B
12. D
13. C

1. B
2. A
3. D
4. C
5. B

1. C
2. D
3. B
4. B
5. A
6. C
7. A
8. D
9. C
10. B
11. A

1. C
2. A
3. B
4. B

1. C
2. B
3. D
4. C
5. False. When the liquid contacts the bottom of the conductance rod or sensor, it activates the switch.
6. False. Bubbler systems are an older technology that has fallen out of favor due to high maintenance costs.
7. A
8. C
9. B
10. True
11. C
12. C
13. A
14. C
15. A
16. True

1. B
2. A
3. D
4. C
5. D
6. True
7. A

8. D
9. B
10. True
11. True

1. True
2. False. Dissolved oxygen does not pass easily through coated membranes.
3. False. Light output decreases with increasing light output.
4. True
5. D
6. A
7. C
8. C
9. B
10. A
11. B
12. C
13. D
14. B
15. D
16. A
17. D
18. B

1. A
2. C
3. C
4. D
5. B
6. D
7. C
8. A
9. B

1. B
2. C
3. D
4. A
5. B
6. C
7. C
8. A
9. B
10. D

1. False. Oxidation–reduction potential readings depend on many factors. Similar probes can give different results.
2. True
3. False. Oxidation–reduction potential readings are difficult to correct for temperature. They do not incorporate temperature sensors.
4. A
5. D
6. B

TOTAL SUSPENDED SOLIDS METERS

1. True
2. True
3. A
4. D
5. B
6. C
7. C
8. D

INTERFACE/SLUDGE BLANKET LEVEL ANALYZERS

1. True
2. True
3. B
4. A

GAS DETECTORS

1. D
2. A
3. C
4. B
5. A
6. C
7. D
8. A
9. False. All sensors will fail. Bump tests help us determine exactly when they need to be replaced.
10. True
11. B

CHAPTER 13

PURPOSE AND FUNCTION AND THEORY OF OPERATION

1. Level 1: Operator
   Level 2: SCADA System
   Level 3: Instrumentation Level
2. Component 1 is a computer.
   Component 2 is networking equipment.
   Component 3 is a controller or PLC.
   Component 4 is an input/output device.
   Component 5 is a field device.

EQUIPMENT

1. False. Supervisory control and data acquisition equipment differs because it is designed to be highly reliable and provide accurate and timely data in an industrial environment.
2. True
3. True
4. False. Only authorized devices should be connected to the facility SCADA network.

OPERATION

1. B
2. A
3. In navigation, "drill down" is the concept of moving between layers of displays (i.e., more general displays at the top and more specific displays at the bottom).
4. Lateral navigation is the idea of moving between displays of the same level without first having to move to a higher-level display. Lateral navigation is important because it reduces the number of clicks needed to navigate between processes.

1. B

CHAPTER 14

PERMITS AND REPORTING

1. False. Maximum Achievable Control Technology 129 relates to the Section 129 requirement for SSI emissions.
2. False. Industrial pretreatment permits are required for nonresidential dischargers who are either a categorical industry or a significant user.
3. False. All WRRFs are required to submit a monthly DMR summary report regardless of any violations.
4. True
5. False. The Clean Water Act regulates point source pollution through the NDPES program. The Safe Drinking Water Act regulates drinking water, not wastewater.
6. True
7. A
8. D

SAMPLING AND RECORDS

1. C
2. B
3. True

BUDGETING AND FINANCIAL MANAGEMENT

1. C
2. True
3. B
4. A

WRITING AND UPDATING STANDARD OPERATING PROCEDURES

1. D

ASSET MANAGEMENT AND MAINTENANCE

1. False. Preventive maintenance is done while the equipment is operating normally.
2. True
3. True
4. False. The failure of critical equipment can be extremely costly. An asset management program can help determine which critical equipment should be replaced before failure and which equipment can be run to failure.

CAPITAL IMPROVEMENTS AND PROJECTS

1. B
2. B
3. C
4. B
5. True

EMERGENCY PLANNING AND RESPONSE

1. C
2. D
3. D
4. C
5. False. Vulnerability assessments consider threats from both human sources and natural disasters—anything that may damage an asset or disrupt operation.
6. False. National Incident Management System and ICS are valuable tools for managing any kind of incident, especially when working with other first responders like the police.
7. D

8. $a = 2, b = 4, c = 3, d = 5, e = 1$
9. C
10. C

PERSONNEL MANAGEMENT

1. A
2. B
3. A
4. B
5. C
6. C
7. B

LEADERSHIP

1. B
2. A
3. D
4. C
5. A

CUSTOMER SERVICE AND PUBLIC ENGAGEMENT

1. False. All complaints should be handled courteously and promptly.
2. D
3. False. The operator should describe the facility in understandable terms.
4. A

Periodic Table

PERIODIC TABLE
group

1	2	3	4	5	6	7	8	9	10	11	12	13	14	15	16	17	18

IA

element name ⟶ **HYDROGEN**
atomic number ⟶ **1**
chemical Symbol ⟶ **H**
atomic weight (u) ⟶ 1,008

VIIIA

period

1
HYDROGEN
1
H
1,008

IIA

Alkali metal	Transition metal	Nonmetal	● Solid
Alkaline earth metal	Post-transition metal	Halogen	● Liquid
Lanthanide	Metalloid	Noble gas	● Gas
Actinide			● Unknown

HELIUM
2
He
4,003

2
LITHIUM
3
Li
6,941
BERYLLIUM
4
Be
9,012

IIIA
BORON
5
B
10,81
CARBON
6
C
12,01
NITROGEN
7
N
14,01
OXYGEN
8
O
16,00
FLUORINE
9
F
19,00
NEON
10
Ne
20,18

3
SODIUM
11
Na
22,99
MAGNESIUM
12
Mg
24,31

IIIB   IVB   VB   VIB   VIIB   VIIIB   VIIIB   VIIIB   IB   IIB

ALUMINIUM
13
Al
26,98
SILICON
14
Si
28,09
PHOSPHORUS
15
P
30,97
SULFUR
16
S
32,07
CHLORINE
17
Cl
35,45
ARGON
18
Ar
39,95

4
POTASSIUM
19
K
39,10
CALCIUM
20
Ca
40,08
SCANDIUM
21
Sc
44,96
TITANIUM
22
Ti
47,87
VANADIUM
23
V
50,94
CHROMIUM
24
Cr
52,00
MANGANESE
25
Mn
54,94
IRON
26
Fe
55,85
COBALT
27
Co
58,93
NICKEL
28
Ni
58,69
COPPER
29
Cu
63,55
ZINC
30
Zn
65,39
GALLIUM
31
Ga
69,72
GERMANIUM
32
Ge
72,59
ARSENIC
33
As
74,92
SELENIUM
34
Se
78,96
BROMINE
35
Br
79,90
KRYPTON
36
Kr
83,80

5
RUBIDIUM
37
Rb
85,47
STRONTIUM
38
Sr
87,62
YTTRIUM
39
Y
88,91
ZIRCONIUM
40
Zr
91,22
NIOBIUM
41
Nb
92,91
MOLYBDENUM
42
Mo
95,94
TECHNETIUM
43
Tc
(98,91)
RUTHENIUM
44
Ru
101,1
RHODIUM
45
Rh
102,9
PALLADIUM
46
Pd
106,4
SILVER
47
Ag
107,9
CADMIUM
48
Cd
112,4
INDIUM
49
In
114,8
TIN
50
Sn
118,7
ANTIMONY
51
Sb
121,8
TELLURIUM
52
Te
127,6
IODINE
53
I
126,9
XENON
54
Xe
131,3

6
CAESIUM
55
Cs
132,9
BARIUM
56
Ba
137,3
LANTHANUM
57
La
138,9
HAFNIUM
72
Hf
178,5
TANTALUM
73
Ta
180,9
TUNGSTEN
74
W
183,9
RHENIUM
75
Re
186,2
OSMIUM
76
Os
190,2
IRIDIUM
77
Ir
192,2
PLATINUM
78
Pt
195,1
GOLD
79
Au
197,0
MERCURY
80
Hg
200,6
THALLIUM
81
Tl
204,4
LEAD
82
Pb
207,2
BISMUTH
83
Bi
209,0
POLONIUM
84
Po
(210,0)
ASTATINE
85
At
(210,0)
RADON
86
Rn
(222,0)

7
FRANCIUM
87
Fr
(223,0)
RADIUM
88
Ra
(226,0)
ACTINIUM
89
Ac
(227,0)
RUTHERFORDIUM
104
Rf

DUBNIUM
105
Db

SEABORGIUM
106
Sg

BOHRIUM
107
Bh

HASSIUM
108
Hs

COPERNICIUM
112
Cn

Lanthanides
CERIUM
58
Ce
140,1
PRASEODYMIUM
59
Pr
140,9
NEODYMIUM
60
Nd
144,2
PROMETHIUM
61
Pm
(144,9)
SAMARIUM
62
Sm
150,4
EUROPIUM
63
Eu
152,0
GADOLINIUM
64
Gd
157,3
TERBIUM
65
Tb
158,9
DYSPROSIUM
66
Dy
162,5
HOLMIUM
67
Ho
164,9
ERBIUM
68
Er
167,3
THULIUM
69
Tm
168,9
YTTERBIUM
70
Yb
173,0
LUTETIUM
71
Lu
175,0

Actinides
THORIUM
90
Th
(232,0)
PROTACTINIUM
91
Pa
(231,0)
URANIUM
92
U
(238,0)
NEPTUNIUM
93
Np
(237,0)
PLUTONIUM
94
Pu
(239,1)
AMERICIUM
95
Am
(243,1)
CURIUM
96
Cm
(247,1)
BERKELIUM
97
Bk
(247,1)
CALIFORNIUM
98
Cf
(252,1)
EINSTEINIUM
99
Es
(252,1)
FERMIUM
100
Fm
(257,1)
MENDELEVIUM
101
Md
(256,1)
NOBELIUM
102
No
(259,1)
LAWRENCIUM
103
Lr
(260,1)

www.ingramcontent.com/pod-product-compliance
Lightning Source LLC
LaVergne TN
LVHW081245270225
804587LV00005B/71